Human-Robot Interaction - Perspectives and Applications

Edited by Ramana Vinjamuri

Published in London, United Kingdom

Human-Robot Interaction - Perspectives and Applications
http://dx.doi.org/10.5772/intechopen.100672
Edited by Ramana Vinjamuri

Contributors
Alejandro Ramirez-Serrano, Parastoo Dastangoo, Ruthber Rodriguez, Roberto Sagaro Zamora, Enrique Marañón, Alexander Alexeis Suarez Leon, Pingguo Huang, Yutaka Ishibashi, Falih Mahdi Salih Alkhafaji, Kazuhiko Terashima, Kazuhiro Funato, Takuyuki Komoda, Sarah Carter, Matthew Studley, Francisco J. Perez-Grau, Pedro U. Lima, Meysam Basiri, Gerhard K. Kraetzschmar, Daniele Nardi, Antidio Viguria Jiménez, Gabriele Ferri, Fausto Ferreira, Sven Schneider, Paul G. Plöger, Deebul Nair, Lun Wang, Emanuele Antonioni, Vincenzo Suriani, Luca Iocchi, Suman Dutta, Ajit Kumar Singh, Bhabani Prasad Mondal, Kiranmoy Patra, Debashis Paul, Seiki Augustine Chiba, Mikio Waki, Tharun Kumar Reddy, Ashok Kumar Chaudhary, Vinay Gupta, Kumar Gaurav, Laxmidhar Behra, Ramana Vinjamuri, Helen Meyerson, Parthan Olikkal, Dingyi Pei

First published in London, United Kingdom, 2023 by IntechOpen
IntechOpen is the global imprint of INTECHOPEN LIMITED, registered in England and Wales, registration number: 11086078, 5 Princes Gate Court, London, SW7 2QJ, United Kingdom

British Library Cataloguing-in-Publication Data
A catalogue record for this book is available from the British Library

Additional hard and PDF copies can be obtained from orders@intechopen.com

Human-Robot Interaction - Perspectives and Applications
Edited by Ramana Vinjamuri
p. cm.
Print ISBN 978-1-80356-410-4
Online ISBN 978-1-80356-411-1
eBook (PDF) ISBN 978-1-80356-412-8

For EU product safety concerns:
IN TECH d.o.o., Prolaz Marije Krucifikse Kozulić 3, 51000 Rijeka, Croatia,
info@intechopen.com or visit our website at intechopen.com.

Meet the editor

Ramana Vinjamuri holds a Ph.D. in Electrical Engineering from the University of Pittsburgh, USA, specializing in dimensionality reduction in control and coordination of the human hand, which he received in 2008. He worked as a postdoctoral research associate in the field of Brain Machine Interfaces (BMI) to control prostheses in the School of Medicine, University of Pittsburgh, during 2008–2012. In 2010, he received the Mary E. Switzer Merit Fellowship from the National Institute on Disability, Independent Living, and Rehabilitation Research (NIDILRR). From 2012 to 2013, he was a research assistant professor in neuroprosthetics at the Department of Biomedical Engineering, Johns Hopkins University, USA. Between 2013 and 2020, he served as an assistant professor in the Department of Biomedical Engineering at Stevens Institute of Technology, USA. In 2018, he was honored with the Harvey N. Davis Distinguished Teaching Award for his excellence in undergraduate and graduate teaching. In 2019, he was the recipient of the National Science Foundation (NSF) CAREER Award and in 2020, he received the NSF IUCRC Center Planning grant. He has also received research awards from NIDILRR, United States-India Science & Technology Endowment Fund (USISTEF), and New Jersey Health Foundation. Currently, he holds a secondary appointment as an adjunct assistant professor at the Indian Institute of Technology, Hyderabad, India. Dr. Vinjamuri is an assistant professor in the Department of Computer Science and Electrical Engineering at UMBC.

Contents

Preface

This book brings together advances and challenges in multiple exciting fields of research including human–machine interfaces, collaborative and humanoid robots, human–robot symbiosis, human–human collaboration, and robotics. Human–robot interaction has significant applications in health care, education, manufacturing, military, space exploration, and much more. This book compiles perspectives and applications in this thriving field of research. From the contributions of renowned international scientists across the world, this book discusses exciting topics such as advances in fundamental robotics and control, deep neural networks in robot grasp recognition, cooperation and adaptation between humans and robots, integration of drones in farming, and current applications in healthcare robots in smart hospitals. This book is a great resource for students, teachers, researchers, engineers, entrepreneurs, and readers interested in the latest developments in human–robot interaction.

The introductory chapter summarizes the advances in collaborative and humanoid robots and applications of human–robot interaction. Chapter 2 discusses EEG control of a robotic wheelchair for individuals with motor disabilities and elderly adults. Chapter 3 presents a classification of gestures in robotic hand grasp by deep neural networks. Chapter 4 examines cooperation among humans and robots in remote robot systems with force feedback. Chapter 5 describes multi-contact humanoid stability for increased interaction in unstructured environments. Chapter 6 discusses materials and sensors needed in human–robot interaction, for example, dielectric elastomer sensors capable of measuring deformation and pressure. Chapter 7 considers the design of the 5 degrees of freedom (DOF) industrial robotic arm. Chapter 8 discusses healthcare robots and smart hospitals based on human–robot interaction. Chapter 9 describes the digital inclusion of the farming sector using drone technologies. Finally, Chapter 10 discusses the European robotics league, which holds smart city robot competitions to encourage these technologies and benchmarking.

I would like to thank the authors for their immensely valuable contributions without which this book would not have been possible. I would also like to thank the staff at IntechOpen, especially Blanka Gugic, Lucija Tomicic-Dromgool, and Sara Tikel for their kind assistance throughout the editing process. I truly hope that readers will benefit from these selected chapters and increase their understanding of human–robot interaction.

Ramana Vinjamuri, Ph.D.
Assistant Professor,
Computer Science and Electrical Engineering,
University of Maryland,
Baltimore County, Baltimore, MD, USA

Chapter 1

Introductory Chapter: Human-Robot Interaction – Advances and Applications

Helen Meyerson, Parthan Olikkal, Dingyi Pei and Ramana Vinjamuri

1. Introduction

Recent advances in robotic technology are bringing about robots better suited to perform tasks and applications in which robots are interacting directly with people in their everyday environments, both at home and in the workplace. Human-robot interaction (HRI) is beneficial because robots have been shown to deliver an emotional response to humans and humans find robots engaging. Additionally, robots can integrate into everyday settings without difficulty and can be perceived by humans as active social agents, meaning they can complete the programmed tasks with total control, independence, and intentionality. With HRI, a user's experience of interaction varies from person to person and is influenced by many factors such as physical context of the environment, cultural context, thoughts and feelings toward the robot, and social nature [1].

HRI is also an important development because it allows robots to be directed by humans to complete certain challenging and hazardous tasks, notably in an industrial setting. With modern computational algorithms programmed into the environment, HRI can increase productivity and reduce downtime and task interruptions [2]. Additionally, HRI is a beneficial solution to compensate for a lack of human labor force in a certain setting, due to various factors such as extreme conditions or low pay. The lack of human labor hurts the local or large-scale economy as it means a lower production supply, and this issue can be resolved potentially by incorporating robots into the scene. However, fully replacing humans with robots would mean a larger initial investment and would eliminate availability of jobs. Instead, robots could be incorporated alongside human workers as a means to improve human comfort and optimize productivity. HRI is a significant modern approach to improve the functioning of everyday settings and has countless advantages and applications.

2. Collaborative and humanoid robots

Collaborative robotics is the field of study that involves using human demonstration to teach robots different skills. The robot can learn to recognize goal-oriented actions and understand human actions and verbal and nonverbal communication. While robots can learn from imitation, in a complex environment where different situations arise, imitation is not enough to make the robot able to function in the complex environment by itself without human involvement. Collaborative robots can work alongside humans on tasks and can provide assistance by responding to user requests for help or by automatically detecting at what point to assist. Thus, in a collaborative environment, both parties must have the ability to refer to objects in the shared space. Humans can use a combination of various techniques, including sensorimotor signals, verbal cues, pointing gestures, and gaze to communicate to the robot to handle a certain object [3].

Humanoid robots are designed to resemble humans in terms of appearance. They have continued to increase their roles in everyday human environments as coworkers, companions, trainers, and assistants. Humanoid robots are created to be similar to humans both in outward design and language and gesture behaviors. In designing robots to play roles alongside humans, it is important to investigate how humans interpret and emotionally respond to the robot to allow for a smooth incorporation into our everyday lives. Humans have been demonstrated to engage with and respond especially well to humanoid robots. Humanoids were seen as having more moral responsibility, observing social norms, and generating formal expressions from their human counterparts communicating with them [4].

Telerobots perform routine tasks under supervisory control by humans. The human supervisors monitor and reprogram the robots at irregular intervals to execute different pieces of the higher-arching task. Telerobots are designed to simplify communication with humans and improve the ease of human control. It is important that the telerobot is directed to complete the task as efficiently as possible while the human operator is comfortable controlling the telerobot, even during chaotic situations. Moreover, they can be instructed to carry out tasks in environments that are hazardous or inaccessible to humans. Additionally, telerobots have greater precision than human hands, which may come useful in many different settings such as surgeries [5].

Figure 1.
Overview of significant applications of human-robot interaction.

Human-robot interaction is an area of research that involves developing and improving the most optimal robots that cooperate with humans. An overview of current and potential applications of HRI is illustrated in **Figure 1**. In the subsequent sections, we discussed each of these applications and challenges in detail.

3. Space exploration

Emerging technologies in HRI look promising to efficiently combine the capabilities of astronauts, remote operators, and robotic assets into human-machine teams that can effectively communicate for the purpose of space exploration. These technologies have been carefully planned to meet sustainability requirements and minimize the use of resources. The use of HRI can be especially beneficial to complete space exploration tasks such as collecting environment and mapping information, providing situational awareness of the scene and surroundings, developing and maintaining infrastructure, and providing mobility support to the astronauts. The future of successful space exploration will be heavily influenced by the ability of the human and robot to demonstrate strong communication through both gestures and dialogue and to collaborate with one another for problem-solving [6].

One such HRI technology is Explainable AI (xAI), which can provide a virtual deep-space environment simulation that can show how the space rover will behave in a certain scenario, and the human controller can prepare strategies and informed decisions to apply during the actual deployment. An additional technology is virtual, augmented, and mixed reality (VAMR), which provides visual displays, situational awareness, and additional functionality and communication. The navigation cues and technology recognition that VAMR provides can guide the rover in effectively investigating an unfamiliar terrain. Another emerging technology is adaptive and adaptable automation. Adaptive control is where the robot automatically adjusts control parameters as a system response, while adaptable control is where the human controller operates manual system changes. This technology is an optimal design that balances the self-adjusting robots and the significance of human monitoring, aiding the efficiency and safety during space exploration [7]. The use of HRI along with the emerging technologies in the area of space exploration expands the possibilities for new learnings and discoveries.

4. Military

The future of military robots puts soldiers and robots as teammates, where the soldier and robot can share the task load and accomplish the goal together. In this environment, the robot is an important entity that acts autonomously and intelligently and can simulate team behaviors such as communication and coordination [8]. Robots are able to complete operations in environments that are harmful to the soldiers, and this keeps the soldiers and civilians safe. These operations include clearing buildings, search and rescue in disaster areas, detecting explosives, and surveillance activities. Additionally, military robots can support the soldiers by gathering data to improve situational awareness, transport equipment, efficiently distribute supplies, facilitate commanders' decision-making, and protect the soldier from hostile attacks.

To make HRI integration possible in this setting, a multitude of factors have to be considered including operating environments, task difficulty, soldier's comfort level with the robot, and communication and decision-making for both the soldier and robot. Emerging technologies, both modeling and simulation systems, have been developed to identify and resolve potential integration issues. One such modeling system is the Improved Performance Research Integration Tool (IMPRINT). IMPRINT analysis demonstrated that integrating HRI into a mission with soldiers mounted in carrier vehicles or on horses would cause overload issues, and, therefore, gunners were a better-suited group for HRI integration. Modeling technologies such as IMPRINT can be used to set guidelines that can be validated through simulations, and the models can be revised and improved. Simulations are particularly helpful to determine the effects of adding complexity to the tasks, considering potential strategies to reduce overload and investigating ways to improve performance while carrying out the military task [9].

5. Healthcare

Robots taking on roles as healthcare workers have incredible benefits for the population. These include accuracy in treatment performance, strong working speed, reducing workload for the human healthcare worker, organization of daily routine, optimizing healthcare resources, and resolving simple problems so that the patient does not have to visit the doctor [10]. The elderly population is increasing in size and the available supply of healthcare workers concerningly cannot support the population increase of this demographic. This demographic especially has the potential to improve well-being as a result of interaction with healthcare robots. Robots as healthcare workers allow elderly adults to be at home later in life instead of in an elder care facility, which reduces financial and emotional stress for the patient and the family, lowers costs, and helps elderly adults retain independence and be happier and healthier.

Healthcare robots can serve in rehabilitation or social roles. Rehabilitation robots can perform tasks or make tasks easier for the user, while social robots are for elderly adults to interact with and have as a companion. It is important to consider the concerns and needs of elderly adults during the robotic design process so that the user will accept the robot. Some elderly adults have demonstrated to be skeptical of accepting the robot due to it being a rapid jump in technology and the potential privacy issues it may present, but they are more likely to accept the robot if it can perform tasks that they find useful [11].

Conditions that the elderly population face that HRI provides technological improvements for include physical and functional decline and cognitive decline. Healthcare robots can assist elderly adults with tasks that become more difficult due to these conditions. Emerging technologies can help with tasks impacted by physical and functional decline such as cleaning, heating food, and sorting laundry. Robotic developments in the areas of mobility assistance and other activities such as bathing have also been in the works. Healthcare robots make these activities safer and more comfortable for the patient. Other technologies help patients monitor their health conditions and provide appointment reminders. Robotic technologies help with cognitive decline by providing cognitive training exercises that keep the patients engaged and stimulated.

The COVID-19 pandemic has only furthered the growing shortage of healthcare workers. Throughout the pandemic, healthcare robots were used for a high variety of purposes, including health screenings, transportation of medical goods, and even direct patient care. The robots provided a multitude of benefits, including minimizing human contact and, therefore, reducing transmission rates and decreasing the workload on healthcare workers. The technologies used during the pandemic were adapted from preexisting technologies, as this approach was more efficient than developing new technologies during a crisis. For example, the Guangzhou Gosuncn Robot Company developed robots originally intended to be used for policing, but these robots were modified and equipped with powerful cameras to screen the body temperature of up to 10 people at once and detect if an individual is wearing a facemask. The COVID-19 pandemic demonstrated how important it is to formulate reliable protocols for how to adapt preexisting technologies for healthcare purposes if and when a future pandemic or crisis occurs [12]. This will allow for the most organized treatment possible and the most efficient patient path to recovery.

6. Manufacturing

HRI provides benefits in manufacturing in terms of productivity, safety, and working conditions. HRI is an approach that complements the strengths of humans and robots in manufacturing. This approach would make manufacturing a more sustainable career for individuals in the long term, as the incorporation of robots allows workers to avoid hard physical work. This also means a reduction in illness rates. Additionally, productivity increases with this approach because robot workers do not need downtime or on-the-job training. HRI can reduce running costs and speed of assembly, and improves readability and precision. Multiple factors need to be considered when implementing HRI into a setting to optimize performance. These include movement speed of the robot, distances between humans and robots, robot noises, trajectory of the robot, and physical appearance of the robot [13].

Robots can especially be helpful in assisting humans in the areas of delivering tools and parts and holding manufacturing equipment objects or objects in the process of being assembled. Robots can contribute accuracy, speed, and consistency to the setting, while humans contribute organization, management, and more cognitive assets. By sharing the workspace, situation awareness, danger perception, and enrichment communication are promoted. Modern robot designs often are programmed with advanced sensing, joint compliance, and artificial intelligence. Robots can play impactful roles in individual parts of the manufacturing setting, or they can contribute to the setting in a broader sense. For example, in a narrower role, the robot can control manufacturing tools or feeder equipment such as conveyors and loaders. In a broader sense, the robot's state-of-the-art design and advanced technology give it the ability to contribute to process flow control and the maintenance of workplace safety.

Manufacturing settings vary in many ways, including plant size, wealth, and typical size of produced batch. HRI is beneficial for industrial settings of all sizes, so it is important to find ways to make HRI more accessible for small and mid-size enterprises (SMEs), which have fewer resources, to begin with, and are less likely to take risks with their manufacturing model [14]. SMEs play a critical role in the economy, and this emphasizes the need for SMEs to adapt to modern technologies so that they can optimize consumerism. In the UK, 99% of the 5.6 million businesses are SMEs.

One approach to increase HRI in SME settings is to identify individual motivation by creating a model that supports fulfilling an overall goal by achieving predefined subgoals. For example, faster and more efficient destocking of assembled parts contributes to greater productivity. The predefined subgoal, in this case, is the faster destocking of assembled parts, which is an area that HRI can demonstrate strong support to. This approach allows the SME to identify the most suitable technologies for the assembly setting without having to use up resources to trial different technologies that may or may not be optimal. HRI has incredible benefits for manufacturing, and it is important to determine the most efficient way to incorporate it into the setting.

7. Education

HRI has shown promise in the area of being learning companions for children in classrooms and at home, and as tutors to help students better understand the content. HRI has been demonstrated to be beneficial for students of all ages, including preschool, elementary school, and post-secondary education. The use of HRI can help teach a broad range of disciplines, including STEM, languages, and handwriting. Aside from adaptability to a wide range of disciplines, robots provide additional benefits in education settings such as engagement, motivation, improving the learner's self-esteem, and providing empathetic feedback. When designing robots for this purpose, it is important for developers to consider the social conscience of the robot and its ability to collaborate with educators.

At the preschool level, emerging technologies are often geared toward improving social integration and engaging the children in constructive learning, meaning the learner is actively involved in knowledge construction. The technologies are incorporated into storytelling in the classroom, as storytelling is essential for children's language and creative development. In this setting, the robot would act as a storyteller to the children. Adopting HRI into storytelling has demonstrated a positive impact on the children's enjoyment and engagement. It has also shown positive results in rehabilitation, learning English, and creativity enhancement [15]. HRI is additionally adaptable to different educational environments, such as a playground or schoolyard, which gives the children further room to learn and grow.

At the elementary school demographic, robots have taken the role of tutors in the area of language learning. In one study, 10- to 11-year-olds were formulated with the task of learning an artificial language. The robots taught the children a 30-minute introductory lesson, and ideally, the students would be able to form simple sentences after the lesson. The sociability of the robot was demonstrated to be a crucial factor in terms of both engagement and performance. The students had stronger engagement and performances when the robot showed role model behavior, personal feedback, empathy, and communicativeness [16]. These findings further support the importance of considering sociability when designing a robot for tutoring purposes.

Personalization has been a recent subject of interest when designing a robot for the educational setting. The extent that tailoring to an individual's strengths and weaknesses is beneficial to that individual's learning, is not fully understood. This subject matter was investigated in a study where undergraduate and graduate student participants were tutored by a robot in solving grid-based logic puzzles. Participants received lessons from both personalized and non-personalized robots. The findings supported that even relatively simple personalization shows significant learning benefits, as personalization led to stronger performance and faster speed of solving the puzzle [17]. This demonstrates that personalization and adaptability is other important quality to keep in mind when designing the most optimal robot for learning. Additionally, it makes clear that HRI is a beneficial approach for post-secondary students and not just for younger students.

8. Personal and societal applications

HRI has emerged in society working with people in airports, shopping malls, and care centers. With robots entering public spaces more often, this comes with the responsibility of having to maintain a positive image and appearance, as well as behavior that reflects well on society. It is critical that robots for these settings are designed to be accepting of all people and not promote gender stereotypes or ageist views. Robots in public settings have the capacity not only to be respectful to those being helped and not show social biases, but they can also go as far as to be an example and advocate for positive social change. They can bring about a positive impact on a wide range of issues in society such as homelessness, poverty, and refugee crises. To develop robots that represent social empowerment, it is important to consider how robots are shaped as part of society's socio-political dynamics.

Airports are one area where the incorporation of HRI can be particularly helpful and improve passenger experience in the setting. Airports are often overwhelming for passengers due to the large crowds, frequent announcements, and confusing screens and signs. The atmosphere of the airport setting should be considered when designing a robot to fulfill the needs of the passengers. A large robot that can communicate using nonverbal gestures is favorable because airports are crowded and noisy and it is important for the robot to be easily accessible and understood despite the surrounding noise. It is also important for the robot to be able to accommodate the hearing impaired, which could be done by having a display space showing text and images. Additional factors to look out for include affordability, range of dynamic motion, and suitability for the particular environment [18].

Retail is a separate area where HRI can benefit customers. HRI can improve service quality by helping customers navigate a store to find products and information, receive personalized guidance on products, order online for delivery or pickup, and complete purchase transactions. As HRI makes it easier for customers to shop, this in turn increases sales, reduces labor costs, and provides an engaging retail experience. Robots in retail additionally have advantages over human staff, as this approach minimizes human error and allows for more rapid service processes. Human staff often experience physical fatigue and mental strain when performing service tasks, and their work experience includes training time and downtime, which takes away from the opportunity for productive sales. Humanoid robots, notably, can mimic human communication and social interactions, and this makes them strong candidates for integration into retail settings. When designing a robot for this setting, it is critical to consider the robot's emotional aspect for an optimal customer experience. The use of HRI in public environments is promising as a means to improve personal experience and have a positive socio-political impact on society as a whole.

9. Challenges in HRI

It is difficult to design a robot that allows for accurate interaction and communication. There is still work to be done to look for creative ways to improve the capacity of robots in understanding human actions and responding appropriately. With the ability to recognize human hand gestures, there is still room for error due to the complexity and high degree of freedom of human hands. More effective robots should combine multi-modal features, and be able to recognize posture, facial expressions, and voice intensity. This comes along with developing more complicated and powerful sensors, which further makes equipping the robot difficult. Additionally, for optimal interaction, robots need a mechanism to foresee and predict upcoming actions. The complexity and inconsistency of human actions make designing this mechanism a challenge. In HRI, robots also need to be sensitive to surroundings, as well as clutter, lighting changes, and depth perception. It is important and at the same time difficult to consider all of these factors together.

HRI not only has certain design challenges, but it also has some ethical parameters. It is important to keep in mind both helpful and harmful behavior with regard to robots and robotic assistance. The use of robots for killing activities in warfare, for sexual pleasure, or to care for emotionally unstable target groups is a particularly sensitive subject matter. Robots also have the potential to make humans less motivated to work, or unwilling or unable to fulfill certain tasks, even simple ones. There are multiple perspectives on robot rights, treating robots respectfully, and if ethics even apply to the robot itself altogether. An additional ethical issue is who regulates robot use, and who is held responsible if a robot causes damage to a human or property. This also brings into question who is responsible for robot malfunctions as well as the proper way to dispose of robots. HRI also has privacy issues, as the process of consenting to give out personal information to the robot is not concrete. Another potential issue is the robot's physical appearance if it is inadvertently built to match any biases of the designer or embody discrimination through having Euro-centric or overly feminized features. HRI has many ethical issues that are important to take into perspective and find ways to avoid possible harm to robots or users.

10. Conclusion

HRI is evidently a promising modern approach with great benefits in both home and work sites. Some of these benefits include providing engagement, accuracy, productivity, and adaptability. Collaborative robots, humanoid robots, and telerobots all have endless possibilities, and there is still improvement space to further explore the promising potential that these technologies offer. When designing a robot for optimal performance, there are many important factors to keep in mind including physical appearance, behavioral traits, and suitability for a particular setting. Emerging technologies, including simulation systems and virtual displays, are helpful in testing and improving a robot's capabilities and preparing for integration. In the subsequent chapters, this book will discuss modern HRI applications in multiple aspects and will touch upon different perspectives and experimental methodologies to develop HRI environments. Emerging technological advancements in HRI and the strong evidence of the incredible benefits make HRI an excellent approach in everyday settings with even more exciting growth to come.

Author details

Helen Meyerson, Parthan Olikkal, Dingyi Pei and Ramana Vinjamuri*
Vinjamuri Lab, University of Maryland Baltimore County, Baltimore, MD, USA

*Address all correspondence to: ramana.vinamuri@gmail.com

References

[1] Young JE, Sung J, Voida A, Sharlin E, Igarashi T, Christensen HI, et al. Evaluating human-robot interaction. International Journal of Social Robotics. 2011;**3**(1):53-67

[2] Landi CT, Ferraguti F, Costi S, Bonfè M, Secchi C. Safety barrier functions for human-robot interaction with industrial manipulators. In: 2019 18th European Control Conference (ECC); 2019. pp. 2565-2570. DOI: 10.23919/ECC.2019.8796235

[3] Kragic D, Gustafson J, Karaoguz H, Jensfelt P, Krug R. Interactive, collaborative robots: Challenges and opportunities. International Joint Conference on Artificial Intelligence IJCAI-18. 2018. pp. 18-25. DOI: 10.24963/ijcai.2018/3

[4] Austermann A, Yamada S, Funakoshi K, Nakano M. How do users interact with a pet-robot and a humanoid. In: CHI'10 Extended Abstracts on Human Factors in Computing Systems Systems (CHI EA '10). New York, NY, USA: Association for Computing Machinery; 2010. pp. 3727-3732. DOI: 10.1145/1753846.1754046

[5] Sheridan TB. Human–robot interaction: Status and challenges. Human Factors. 2016;**58**(4):525-532

[6] Arora A, Panda SN, Raheja J, Nagpal D. Development Approaches To Intuitive, SSD & Haptics Integrated HRI & Social HRI systems for Assisting Space Exploration. In: 2021 6th International Conference on Innovative Technology in Intelligent System and Industrial Applications (CITISIA); 2021. pp. 1-7. DOI: 10.1109/CITISIA53721.2021.9719922

[7] Luebbers M, Chang C, Tabrez A, Dixon J, Hayes B. Emerging Autonomy Solutions for Human and Robotic Deep Space Exploration. In: Proceedings of Space CHI: Human-Computer Interaction for Space Exploration (SpaceCHI 2021). Yokohama, Japan. 2021

[8] Demir M, McNeese NJ, Cooke NJ, Ball JT, Myers C, Frieman M. Synthetic teammate communication and coordination with humans. In: Proceedings of the Human Factors and Ergonomics Society Annual Meeting. Los Angeles, CA: SAGE Publications; 2015

[9] Cosenzo KA, Barnes MJ. Human-robot interaction research for current and future military applications: From the laboratory to the field. In: Proc. SPIE 7692, Unmanned Systems Technology XII. Vol. 7692. 7 May 2010. DOI: 10.1117/12.850038

[10] Broadbent E, Kuo IH, Lee YI, Rabindran J, Kerse N, Stafford R, et al. Attitudes and reactions to a healthcare robot. Telemedicine and e-Health. 2010;**16**(5):608-613

[11] Broadbent E, Tamagawa R, Patience A, Knock B, Kerse N, Day K, et al. Attitudes towards health-care robots in a retirement village. Australasian Journal on Ageing. 2012;**31**(2):115-120

[12] Zhao Z, Ma Y, Mushtaq A, Rajper AMA, Shehab M, Heybourne A, et al. Applications of robotics, artificial intelligence, and digital technologies during COVID-19: A review. Disaster Medicine and Public Health Preparedness. 2021;**2021**:1-11

[13] Bortot D, Ding H, Antonopolous A, Bengler K. Human motion behavior while interacting with an industrial

robot. Work. 2012;**41**(Supplement. 1):1699-1707

[14] Schönfuß B, McFarlane D, Athanassopoulou N, Salter L, Silva LD, Ratchev S. Prioritising low cost digital solutions required by manufacturing SMEs: A shoestring approach. In: International Workshop on Service Orientation in Holonic and Multi-Agent Manufacturing. Cham: Springer; 2019. pp. 290-300

[15] Fridin M. Storytelling by a kindergarten social assistive robot: A tool for constructive learning in preschool education. Computers & Education. 2014;**70**:53-64

[16] Saerbeck M, Schut T, Bartneck C, Janse MD. Expressive robots in education: Varying the degree of social supportive behavior of a robotic tutor. In: Proceedings of the SIGCHI Conference on Human Factors in Computing Systems (CHI '10). New York, USA: Association for Computing Machinery; 2010. pp. 1613-1622. DOI: 10.1145/1753326.1753567

[17] Leyzberg D, Spaulding S, Scassellati B. Personalizing robot tutors to individuals' learning differences. In: 2014 9th ACM/IEEE International Conference on Human-Robot Interaction (HRI). Bielefeld, Germany. 2014. pp. 423-430

[18] Tonkin M, Vitale J, Herse S, Williams MA, Judge W, Wang X. Design methodology for the UX of HRI: A field study of a commercial social robot at an airport. In: Proceedings of the 2018 ACM/IEEE International Conference on Human-Robot Interaction (HRI '18). New York, USA: Association for Computing Machinery; 2018. pp. 407-415. DOI: 10.1145/3171221.3171270

Chapter 2

EEG Control of a Robotic Wheelchair

Ashok Kumar Chaudhary, Vinay Gupta, Kumar Gaurav, Tharun Kumar Reddy and Laxmidhar Behera

Abstract

The Brain-Computer Interface (BCI) technology has been widely used in clinical research; however, its adoption in consumer devices has been hindered by high costs, poor reliability and limited autonomy. In this study, we introduce a low-cost, open-source hardware-based, consumer-grade product that brings BCI technologies closer to the elderly and motor-impaired individuals. Specifically, we developed an autonomous motorized wheelchair with BCI-based input capabilities. The system employs the ROS-backend navigation stack, which integrates RTAB-MAP for mapping, localization, and visual odometry, as well as A* global and DWA local path planning algorithms for seamless indoor autonomous operations. Data acquisition is accomplished using OpenBCI 16-channel EEG sensors, while Ensemble-Subspace KNN machine learning model is utilized for intent prediction, particularly goal selection. The system offers active obstacle avoidance and mapping in all environments, while a hybrid BCI Motor Imagery based control is implemented in a known mapped environment. This prototype offers remarkable autonomy while ensuring user safety and granting unparalleled independent mobility to the motor-impaired and elderly.

Keywords: brain-computer interface, motor-impaired, wheelchair, ensemble-subspace KNN

1. Introduction

The development of Brain-computer Interface (BCI) technology has led to a wide range of scientific and practical applications since its inception in the 1970s. One of the key areas of focus for BCI technology is in the field of wheelchair systems, where the ease of use and efficiency for the user is of paramount importance. This means that the system should be designed in such a way that it is simple for the user to operate the wheelchair and achieve their desired objectives. In particular, EEG-based brain-computer interfaces (BCI) are particularly well-suited for this application as they offer a high degree of convenience and efficiency for the user. However, previous wheelchair systems based on the P300 [1, 2] and those based on steady state visual evoked potentials (SSVEP) [3] did not provide the user with the same level of convenience because they required them to continuously watch a screen in order to decipher commands from EEG signals. The user's field of vision is restricted and fixed on the BCI feedback, making it difficult for them to handle some situations. A wheelchair system based on motor imagery has been suggested in this situation [4, 5]. One of the wheelchair systems has been introduced with fixed direction steering of the wheelchair [6]. But had several shortcomings i.e. one of which was the design approach. It was a non-autonomous and requires constant attention. It does not have synchronous localization and mapping (SLAM) feature. It suffers low information transfer rate. So, to overcome the above shortcoming LiDAR based SLAM was introduced [7]. It is based on steady state visually evoked potentials (SSVEP). But the shortcomings were with shared autonomy and also suffers low information transfer rate and cost ineffective too. A new wheelchair system utilizing motor imagery [8]. was recently introduced, offering a significant enhancement in functionality compared to previous models. However, even more advanced technology was later developed, incorporating both motor imagery and P300 technology to provide even greater enhancement [9] in performance. This system offers the added feature of laser range finder and encoder-based localization, as well as autonomous capabilities. Despite these advancements, the system still lacks the ability to perform inferior 2D mapping and localization, and is cost-ineffective. Additionally, a SSVEP-based direction and angle control system for wheelchair design [10] has been proposed, utilizing visual landmarks for feedback. However, this system is not fully autonomous and also lacks the ability to perform inferior mapping and localization techniques. Additionally, it is cost-ineffective due to its low information transfer rate. Overall, while there have been significant advancements in wheelchair technology, there is still room for improvement in terms of autonomy and cost-effectiveness. A wheelchair based on eye-blink steering control is proposed [11]. An innovative approach but does not solve the issue natural eyeblink signals. Since it is non- autonomous, higher magnitude natural eyeblink signals leads to undesirable motor control of the wheelchair. To overcome this, eye-blinking method is integrated with electroencephalogram (EEG) to control wheelchair movement [12], but sometimes it becomes very difficult and for a person to continuously blink their eyes. As the technology progress, a wheelchair was designed to drive in four directions [13]. It was mainly proposed using time-frequency domain analysis of EEG signals using Neural Network. It is only a proposed prototype, not a full-size wheelchair and not suitable for actual real time implementation. Then a recent wheelchair is introduced based on computer vision-based navigation [14]. It uses tags to localize itself incorporating computer vision, but suffers low maneuverability and unable to handle dynamics obstacles accurately in real time. An omni-directional moveable wheelchair design is proposed based on Mecanum wheel [15]. It is based on SSVEP and Alpha-wave based asynchronous control but the major disadvantage is non-autonomous and requires constant attention for the control mechanism.

The literature reviewed above presents a comprehensive comparison of the various approaches that have been proposed for brain-controlled robotic wheelchairs. It is evident that most of the earlier methods relied heavily on Motor Imagery or P300 for classifying navigation commands. A significant number of researchers employed SSVEP to select from a fixed set of predefined commands. Additionally, some studies captured EEG signals corresponding to eye blinks for prediction purposes. However, when it comes to the chair's autonomy, most early researchers attempted to limit themselves to a fixed set of controls, while others attempted fully autonomous navigation. However, the techniques they used were not as advanced as those available today, making them unsuitable for current use. The proposed method, on the other hand, has successfully overcome the challenges of signal acquisition, goal prediction, mapping, localization, and autonomous navigation. In conclusion, the proposed method is far superior to earlier approaches in terms of its ability to effectively control a brain-controlled robotic wheelchair.

The proposed method for design, development of a bio-signal enabled robotic wheelchair for motor-disabled and elderly care includes following novelties over other existing methods.

1. Proposed design of a low-cost bio-signal enabled robotic wheelchair is with close attention to hardware and electronic safety features, designed specifically for human use.

2. The wheelchair's built-in control system enables rapid and precise point-to-point positioning with little manual control through joystick/directional pad.

3. Maps of the wheelchair environment can be created using the newly introduced mapping module, which can also test working memory for matching point clouds. The wheelchair can localize itself and retain extremely precise odometry of its movement once it has been identified or mapped.

4. Through manual objective selection made available by a GUI touch screen module, point-to-point path planning is feasible in all realized or imported maps.

5. Selected endpoints will be identified as accessible goals within known surroundings for the user to choose via BCI-based control.

6. With these extra features over a standard motorized wheelchair, the elderly and others with motor impairments can independently do daily duties and participate in social activities.

2. Proposed methodology for bio-signal enabled control system

To achieve our objective, we have devised a two-part solution. The first part involves acquiring the goal state through a GUI-based display module (**Figure 1**).

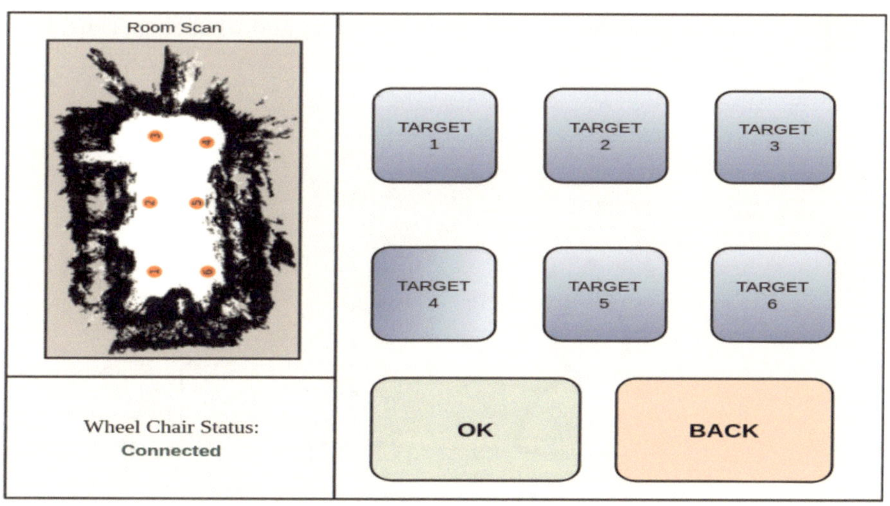

Figure 1.
Graphical user interface (GUI) of wheelchair display.

This module displays a map obtained from the working memory and provides the user with various goal point options. For instance, in a hospital scenario, these options may include patient rooms, gardens, toilets, cafeterias, etc. The user selects a goal point, which triggers a neural depolarization pattern within the cortex. We use a 16-channel OpenBCI cap to record this pattern. By analyzing the time-series data generated by this pattern, we use Motor Imagery (MI)-based prediction and machine learning to estimate the probability of reaching the selected goal.

The navigation module is responsible for devising a trajectory towards the target state while ensuring the safety of the user and avoiding any potential static or dynamic obstacles. Initially, the module focuses on obtaining drift-free odometry by employing a multi-sensor system, including LiDAR, depth camera, IMU, and wheel odometry data that pass through an Extended Kalman Filter (EKF) to ensure accurate odometry. Subsequently, RTAB-Map maintains a live map within the Odom frame of the robot to enable obstacle avoidance via the local planner. The global planner formulates an optimal path from the start state to the goal state based on the static map. Subsequently, the local planner leverages this live map and the navigation waypoints to maneuver in the immediate vicinity while considering the kinematic constraints of the wheelchair, preserving a pre-set safety buffer, and accounting for the motion of moving obstacles. The local planner constantly updates the path to the goal, adhering to imposed environmental and kinematic constraints to ensure complete safety for the occupant until they reach their destination. **Figure 2** shows all the components for bio-signal enabled control system.

2.1 Hardware control module (HCM)

The hardware infrastructure of a brain-computer interface (BCI) wheelchair is a critical determinant of its operational and practical efficiency. It encompasses the sensors and electrodes that capture brain signals, the control mechanism that deciphers these signals and translates them into motion instructions, and the actuators that execute the movement of the wheelchair. The absence of a properly functioning hardware infrastructure could impede the BCI wheelchair's ability to respond to user input and hinder its mobility capabilities. Furthermore, the hardware infrastructure must be robust and dependable to guarantee user safety and wellbeing. Thus, it is

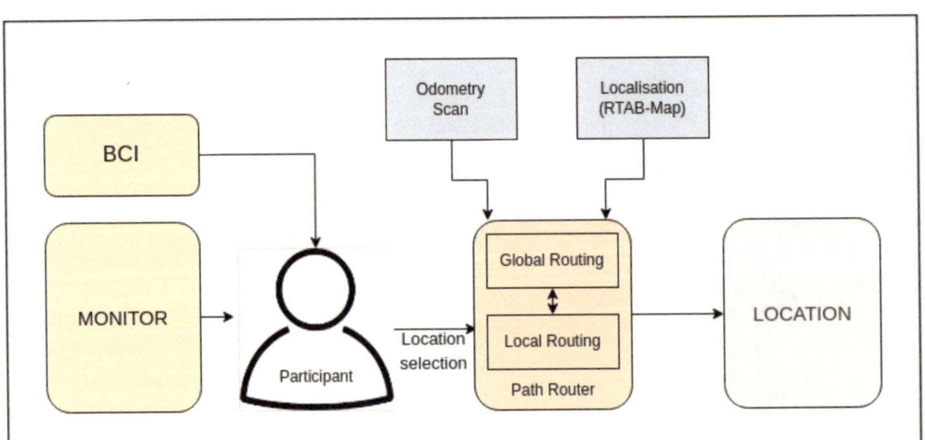

Figure 2.
Block diagram for bio-signal enabled (BSE) control system.

essential to utilize high-quality components and to regularly service and upgrade the hardware infrastructure to maintain its ongoing functionality.

2.1.1 Frame dimension and sensor

The wheelchair has a frame made of iron, with a dimension of 114×64×93.5 cm. It is equipped with two motors and a differential drive system, with a wheelbase of 13 inches in diameter. The maximum weight it can carry is 100 kilograms, and it weighs 45 kilograms without any sensors. The wheelchair has a battery capacity of 24 V, and it can travel up to 20 km on a full charge. Overall, these hardware specifications make the wheelchair sturdy, reliable, and suitable for providing assistance to individuals with mobility impairments. The wheelchair has been modified in order to house the following sensors shown in **Figure 3**.

1. **LiDAR Sensor:** A 2D Light Detection and Ranging (LiDAR) sensor, **Slamtec RPLIDARA1** is being used to get the approximate location information of static and dynamic obstacles with respect to the wheelchair. It has a range of 12 m with a depth resolution of 0.1 mm. The sampling frequency is 4000 Hz at normal mode and 8000 Hz at boost mode. The horizontal field of view is 360°.

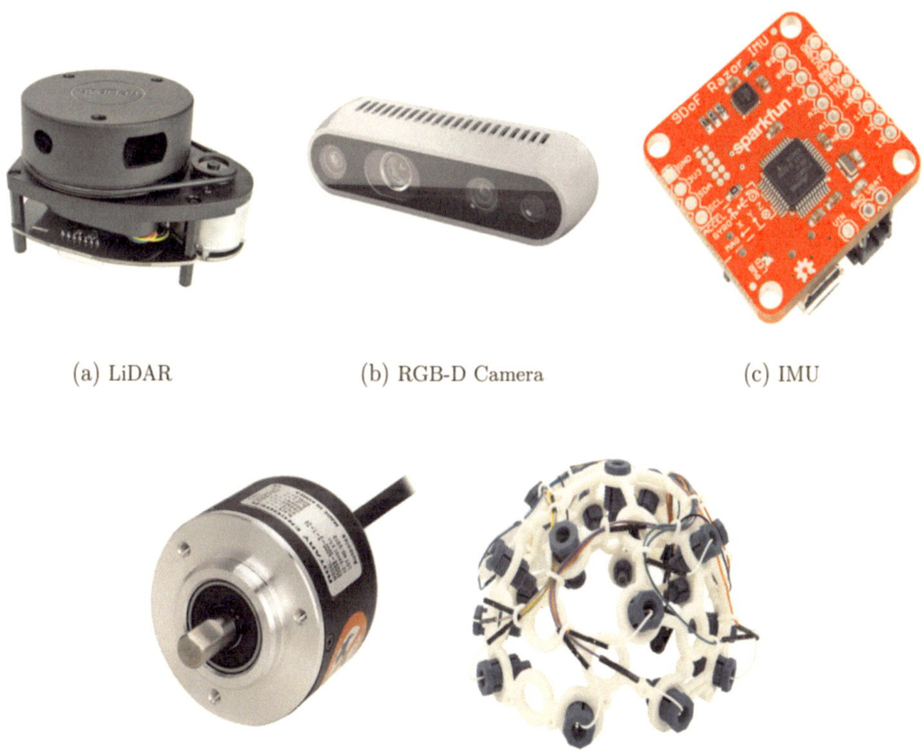

(a) LiDAR (b) RGB-D Camera (c) IMU

(d) Encoder (e) EEG Headset

Figure 3.
Various electronic sensors: (a) LiDAR, (b) RGB-D camera, (c) IMU (d) encoder and (e) EEG headset.

2. **RGB-D Camera:** The wheelchair is equipped with an RGB-D Camera, **Intel Realsense D435i** for getting a 3D point cloud of the environment. RGB-D camera is being used over a 3D LiDAR for getting the point cloud because an equivalent 3D lidar cost 20 times more. It has a resolution of 1280×720 pixels at 30 frames per second with an operating range of 0.11–10 m. The horizontal depth field of view is 85.2°, and the verticle is 58°.

3. **IMU:** SparkFun 9DoF Razor IMU M0 is used as Inertial Measurement Unit. The MPU-9250 in the 9DoF Razor is equipped with three 3-axis sensors: an accelerometer, a gyroscope, and a magnetometer, which allow it to detect linear acceleration, angular rotation velocity, and magnetic field vectors.

4. **Encoder:** Autonics' E30S4 Incremental Rotatory Encoder is used to get the motor feedback. Max response frequency is 300 kHz and max. Allowable revolution is 5000 rpm.

5. **BCI Cap:** An OPENBCI Ultracortex mark 4 EEG headset is used for signal acquisition, and an OPENBCI Cyton biosensing board to sample and feed the EEG signals over Bluetooth to the main computing device. It is capable of collecting up to 16 channels of EEG data from up to 35 distinct 10–20 sites.

2.1.2 Electronics control module (ECM)

- **Motors:** We are using two Robodo MY1016ZL 250 W PMDC motors as the primary drive motors for the wheelchair. Its operating voltage is 24 V, rated torque is 12.7 Nm, and rated speed is 120 RPM.

- **Microcontroller:** Arduino Mega is used as a microcontroller board to control motors and get encoder data. It uses a powerful and power-efficient 8-bit chip, ATmega2560, capable of running instruction at 16 MHz. It has 100 GPIO pins, 4-UART, 5-SPI, I^2C for interfacing various sensors and actuators, connected to the network through USB-Serial.

- **Motor Driver:** To control the speed and the direction of the brushed DC motors, a motor driver of the appropriate specification is needed. We are using Cytron's MDDS30 Smart Drive, which accepts the PWM control signals from the microcontroller. It is rated for motors having a rated current of 30 A, the peak current of 80 A with operating voltage from 7-35VDC. It also comes with regenerative braking technology to charge the battery when one applies the brake.

- **Power Distribution Module:** Different sensors and peripherals need different voltage values and have additional current requirements. Jetson Nano works on 5 V with a peak current of 4 A, while Intel NUC requires 19.5 V. We developed a power distribution module that can provide the stable and required power for all the components to facilitate this power distribution module that can provide the stable and required power for all the components to facilitate this.

- **Display:** To output different destination goals and feedback, we incorporated a 7-inch HDMI touch display connected to the NUC.

DOI: http://dx.doi.org/10.5772/intechopen.110679

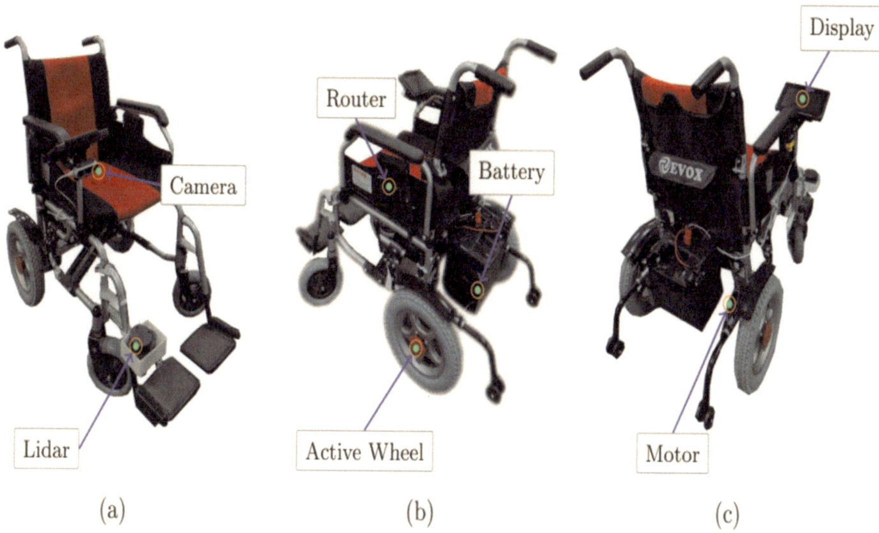

Figure 4.
Various electronics component placement.

The electronics components depicted in **Figure 4** include state-of-the-art sensors and controllers that are designed to respond to the unique needs of the elderly disabled population. These devices work together seamlessly to provide the necessary support and assistance required for a range of mobility tasks.

2.2 Communication control module (CCM)

The system has been designed with a modular approach to actively distribute workloads among sensors, actuators, display devices, and compute modules. This design not only makes the system easy to troubleshoot and repair, but also allows for efficient communication among these components. The **Figure 5** provided illustrates the overall sensor interface and network communication infrastructure that has been implemented. Furthermore, the modular design of the system allows for effective load balancing, making it more reliable and efficient. Overall, the system is designed in a way to make it simple to debug and repair, and the sensor interface and network communication infrastructure are clearly illustrated in the given **Figure 5**, allowing for easy understanding of how the different components communicate with one another. The display flashes probable destinations where the wheelchair can navigate. The destination goal is captured from the brain by the BCI headset. It transfers this information to the Intel NUC wirelessly through Bluetooth, where path planning algorithms use it to plan the path. The feedback from sensor data fusion from the camera attached to NUC, LiDAR and IMU hooked to Jetson Nano 2 and encoders attached to Jetson Nano 1 are used by the system to generate forward and angular velocity for the wheelchair. These velocities are sent to the microcontroller connected to Jetson Nano 1 via serial interface, converting to individual wheel velocities set by the motor driver. All three computing devices are connected through a router via ethernet. **Figure 5** shows the sensor interface and network communication module.

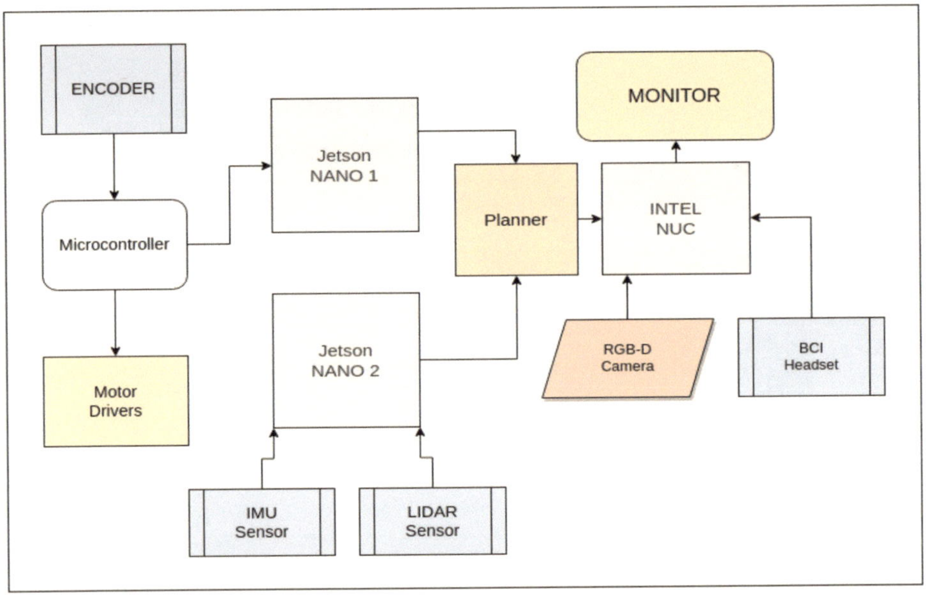

Figure 5.
Sensor interface and network communication.

3. Movement control system

The motor control design for a brain control wheelchair involves using signals from the brain to control the movement of the wheelchair. It includes safety features such as obstacle detection and emergency stop mechanisms. The control algorithms used to interpret the brain signals and control the motors are crucial for smooth and responsive movement. The design also focuses on making the wheelchair easy to operate and user-friendly.

3.1 Wheelchair movement analysis

The robot has two motor controlled wheels at the back and two castors in the front. The two motor controlled wheels control the kinematics (**Figure 6**) of the chair which is implemented using differential drive kinematics. For defining this we take into account several variables. Where, $I_{CC:}$ Instantaneous centre of curvature; R: Radius of curvature; v_l : Velocity of left wheel; v_r : Velocity of right wheel; v_f : Linear velocity of the bot; w_l : Angular velocity of left wheel; w_r : Angular velocity of right wheel; w: Angular velocity of the bot; D_w : Distance between the left and right wheel; d_w : Radius of the wheels.

$$v_f = \frac{v_l + v_r}{2} \tag{1}$$

$$w = \frac{v_l - v_r}{D_w} \tag{2}$$

$$R = \frac{v_l + v_r}{2(v_l - v_r)} \tag{3}$$

DOI: http://dx.doi.org/10.5772/intechopen.110679

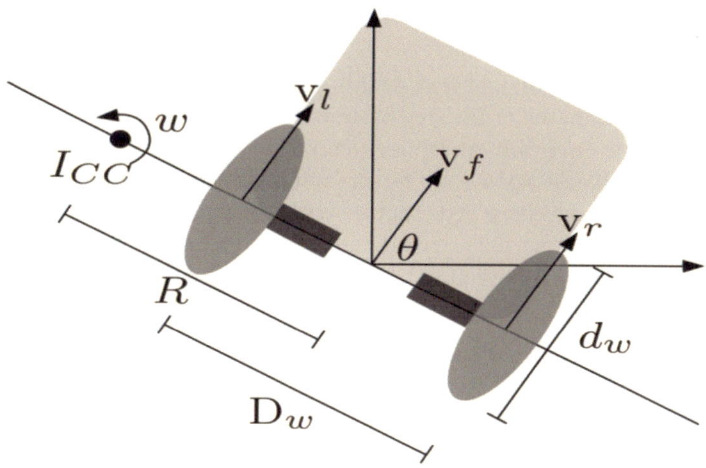

Figure 6.
Wheelchair kinematic model.

$$v_l = d_w * w_l \quad (4)$$

$$v_r = d_w * w_r \quad (5)$$

3.2 PWM based control mechanism

Once the wheel angular velocities have been determined, motor control is achieved using our control system, which consists of a microcontroller connected to one of the Jetson Nano processors via USB. The microcontroller communicates with the motor drivers using Pulse Width Modulation (PWM) signals. They have a PWM pin and an enable pin, the first determines motor speed, while the other is used to determine motor direction. Based on the input from these two pins, delivered via a microcontroller, the motor driver regulates current delivered to the motor from the main power supply battery. The frequency and width of the PWM pulses determine the speed of rotation of the motors. Here, v_{max} and v_{min} have been defined for safety as the maximum and minimum velocities for normal operation.

$$w_l = \frac{2\pi P_l}{P_n} \quad (6)$$

P_l– Number of pulses sent to the left wheel in 1 sec.
P_m– Number of pulses needed for one complete revolution of the wheel.

3.3 Velocity communication

The computer publishes the reference speed commands as wist message on the cmd_{vel} topic. Then, the microcontroller reads the cmd_{vel} from usb serial and generates the PWM signals for motor control. The required linear velocity v_f is in *msg.linear.x* and the required angular velocity w is in *msg.angular.z*. Now solving the Eqs. (1) and (2) the required individual velocities of the wheels are found and then from six the required PWM values are calculated.

4. Autonomous navigation

Simultaneous localization and map building (SLAM) [16] and path-planning are at the core of any autonomous or assisted system. In order to provide for reliable locomotion in a dynamic environment upto a determined goal, we have designed a robust navigation stack. The primary step in navigation is determination of the environment. This is followed by pinpointing ego position within the realized environment - and its update as it relocates within the environment. Any dynamic obstacles must be detected in this process and added to the point cloud. Finally, using the determined map, the procured goal and a constantly updated laser scan, we can navigate as required.

4.1 Various SLAM approaches

We can also objectively evaluate functionality of common slam algorithms on the basis of input/output capabilities. **Figure 7** shows comparison of various SLAM algorithms.

- **Gmapping** [17]: It is the native SLAM approach implemented out of the box in ROS. It is a particle filter based approach that maps 2-D lidar scan to a 2-D costmap. This cost-map is available online, along with pose, and has been widely adapted in conjunction with *amcl*.

- **Hector SLAM** [18]: It is a significant improvement over the above particle filter approach, and can accept imu input for odometery, however external odometery estimates are not accepted by Hector SLAM. Moreover, there is no loop closure support in either of these algorithms, thus earlier estimates of SLAM are not updated upon return.

- **Google Cartographer** [19]: It is one of the earliest open-source SLAM implementations in ROS to implement loop closure detection, thus constantly refining mapping estimates. It use lidar graph to generate a 2-D or 3-D map from laser scans of 2-D or 3-D LiDAR respectively.

- **ORB SLAM2** [20]: It is an improvement on the above graph based approach, and is also a visual SLAM approach, accepting both depth and stereo cameras as inputs.

Algorithm	Depth Cam	Stereo Cam	IMU	LiDAR	Odom	Pose	Occupancy Map 2D	Occupancy Map 3D
GMapping				✓		✓	✓	
Hector-SLAM				✓			✓	✓
Cartographer				✓	✓	✓	✓	✓
ORB-SLAM2	✓	✓						
RGBD-SLAM2	✓		✓		✓	✓		✓
RTAB-Map	✓	✓	✓	✓	✓	✓	✓	✓

Figure 7.
Comparison of various SLAM algorithms.

- **RGBD-SLAMv2** [21]: It is another visual SLAM implementation that can handle full occlusion and white noise added to visual data-stream. It compensates loss of data in such situations by multi-sensor fusion with IMU data, implemented using robot localization package in ROS.

5. Integration of BCI with wheel chair

Goal is selected from the display GUI (**Figure 1**) using BCI module. Once the BCI module gives the final prediction of goal, the autonomous path planning modules kicks in. The Global planner, A* generates an efficient path according to the goal, and the DWA local planner ensures any dynamic change in environment and obstacle avoidance. Motion control system sets the velocities of wheels and odometry from the sensor fusion and localisation give feedback on pose and velocities. This loop continues until the wheelchair reaches its destination. If there is any fallback in any of the modules, recovery behaviors take control for the user's safety. After reaching the destination goal, it waits for the next goal. **Figure 8** shows the BCI integration with wheel chair.

5.1 Brain-computer interface (BCI)

Brain-Computer Interface (BCI) creates an interface between our brain and the computer. We are able to do this because we get different and differentiable signals for every task we do. We analyze these signals and translate them into commands that are sent to an output device to perform a desired action. In our case, we are using these signals to drive a wheelchair.

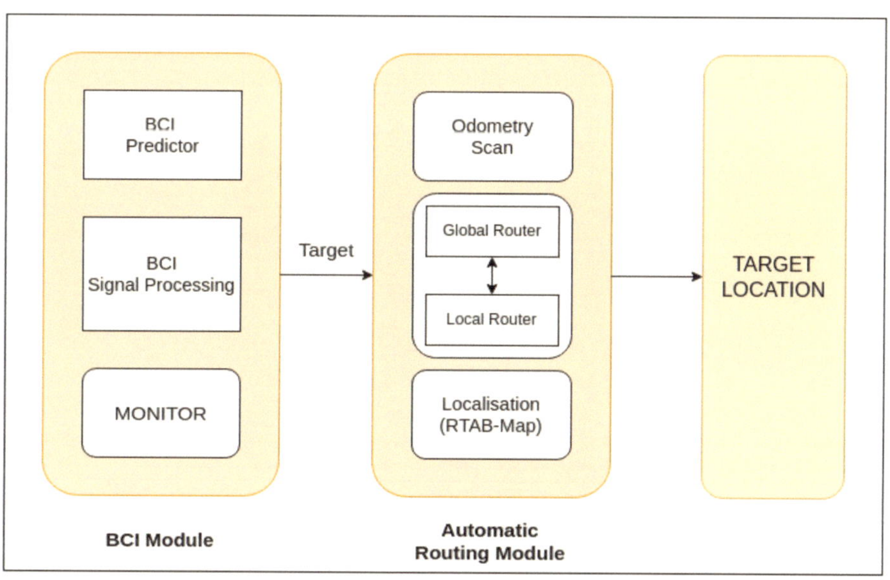

Figure 8.
BCI integration with wheel chair.

5.2 Motor imagery (MI)

MI is one of the standard techniques in BCI in which user is asked to imagine motor action like raising left hand or right limb without actually performing that action. This translates to potential drop that is captured by a EEG Headset. When a person imagining process, event related synchronization and de-synchronization occurs which lie in the frequency range between Mu/Alpha (8 12 Hz) and Beta (16 25 Hz) [22].

5.3 Collection and initial processing of bio-signal recordings

We take the signals from a 16-channel OPENBCI Ultracortex Mark-4 EEG headset (**Figure 3e**), which takes signals from the brain and wirelessly sends them to the computing device using Bluetooth. The plavement of electrodes is very important for motor imagery applications. Standard 10–20 electrode placement can't be found in the **Figure 9**. We have exploited C3, C4, CP1 and CP2 because they provide better signals for motor imagery applications.

Once we get the raw data, we filter it to get away with noise and artifacts. Artifacts are needed to be removed before feeding this data for feature extraction otherwise they interfare with signal of interest which decreases the Signal to Noise Ratio (SNR). Some of them are [23]:

- *ECG Artifacts*: Arises due to heart and pulsing.
- *Muscular Artifacts (MA)*: Arise due to muscular movement.
- *Ocular Artifacts (OA)*: Arise due to eye blinking and movement.
- *Electrode Contact Artifacts*: Arise due to improper electrode contact and due to formation of salt bridge between scalp and electrode due to sweat.

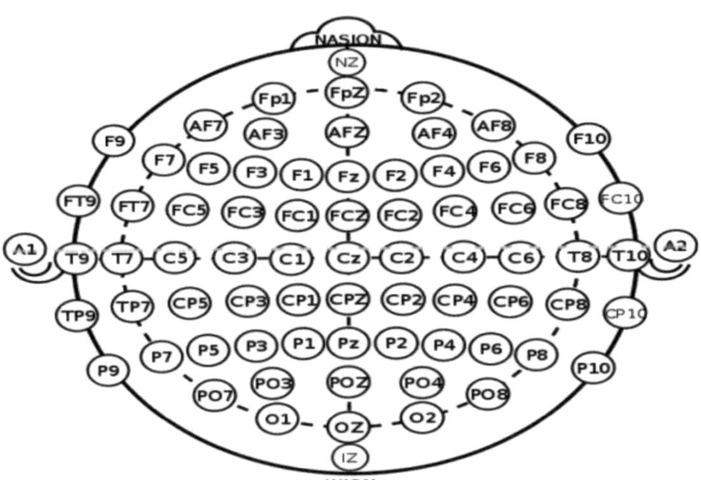

Figure 9.
EEG electrode placement [Source: *commons.wikimedia.org*].

The frequency required for our operation is from 8 to 50 Hz, so we put a fifth-order bandpass butterworth filter of 4 to 60 Hz to remove unwanted frequencies. This bandpass filter also removes some of the low-order frequencies which arise due to head and body movements.

5.4 Data collection sequence

For Motor Imagery-based data collection, we divided the training sequence into four classes: left hand, right hand, left feet, and right feet (**Figure 10**). A blank screen was followed by one of the four sequences flashed on display. EEG data of 5 seconds each were collected, of which we trim 30 sec from start and end and divided into four parts. Slicing allows us to collect data efficiently.

5.5 Feature extraction

Once we have filtered data, we operate on each point and extract features for our machine learning model to classify them into different classes. Features that we have used are as follows:

- Mean,

$$\overline{x} = \frac{\sum_{i=1}^{N} x_i}{N} \tag{7}$$

Median, med(x)
Root Mean Square,

$$RMS(x) = \sqrt{\frac{1}{N}\sum_{i=1}^{N} x_i^2} \tag{8}$$

Variance,

$$\text{Vax}(x) = \frac{\sum_{i=1}^{N} (x_i - \overline{x})^2}{N-1} \tag{9}$$

Skewness,

$$\text{Skew}(x) = \frac{\sum_{i=1}^{N} (x_i - \overline{x})^3}{(N-1) * \sigma^3} \tag{10}$$

where σ is standard deviation.

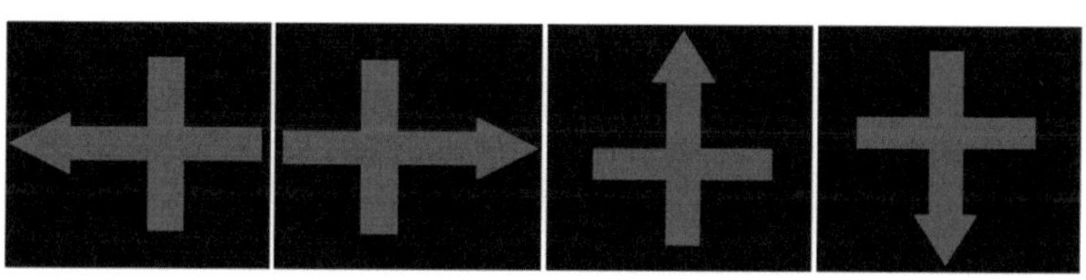

(a) Left Hand (b) Right Hand (c) Right Feet (d) Left Feet

Figure 10.
Motor imagery training sequence.

Kurtosis,

$$\text{kurt}(x) = E\left[\left(\frac{X-\mu}{\sigma}\right)^4\right] \tag{11}$$

Integral features like area under the curve and waveform length.
Slope sign change,

$$\text{Slope Sign Change} = \sum_{n=2}^{N-1} [f[(x_n - x_{n-1})x(x_n - x_{n-1})]] \tag{12}$$

where,

$$f(x) = \begin{cases} 1, & \text{if } x \geq \text{threshold} \\ 0, & \text{otherwise} \end{cases}$$

Apart from these time-domain features, we have also used frequency domain features. But to transfer time-domain EEG signals to frequency domain, we perform a discrete Fast Fourier Transform (FFT) [24] on time domain signal:

$$x[k] = \sum_{n=0}^{N-1} x[n]e^{\frac{-j2\pi kn}{N}} \tag{13}$$

Mean frequency,

$$f_{\text{mean}} = \frac{\sum_{i=0}^{N} I_i \cdot f_i}{\sum_{i=0}^{N} I_i} \tag{14}$$

where, I is spectrogram intensity (in dB)
Median frequency, median(f)
Spectral Power Density [25]

$$x_i(n) = x(n+iD), \quad n = 0,1,2,\ldots,M-1 \tag{15}$$

while; i = 0, 1, 2, ... , L-1;

$$\overset{\approx(i)}{P}_{xx}(f) = \frac{1}{MU}\left|\sum_{n=0}^{M-1} x_i(n)w(n)e^{-j2\pi fn}\right|^2. \tag{16}$$

$$P_{xx}^{W} = \frac{1}{L}\sum_{i=0}^{L-1} \overset{\approx(i)}{P}_{xx}(f). \tag{17}$$

Peak frequency, frequency corresponding to peak of spectral power distribution.

5.6 Training ANS classification

We show one of the four MI training sequence images in a training routine and ask user to imagine those. The EEG headset takes this data and wirelessly send this for

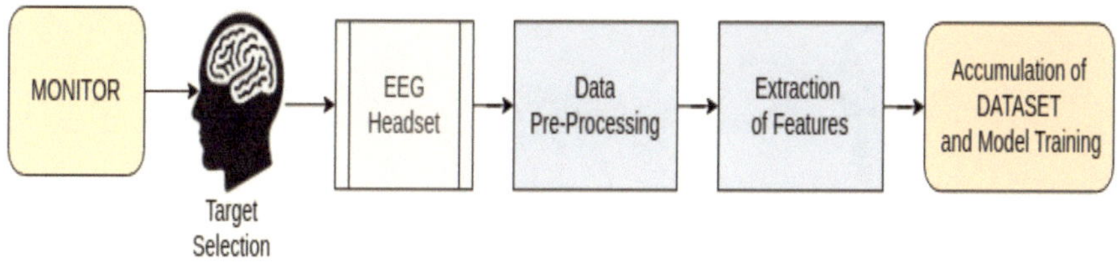

Figure 11.
Block diagram for BCI training module.

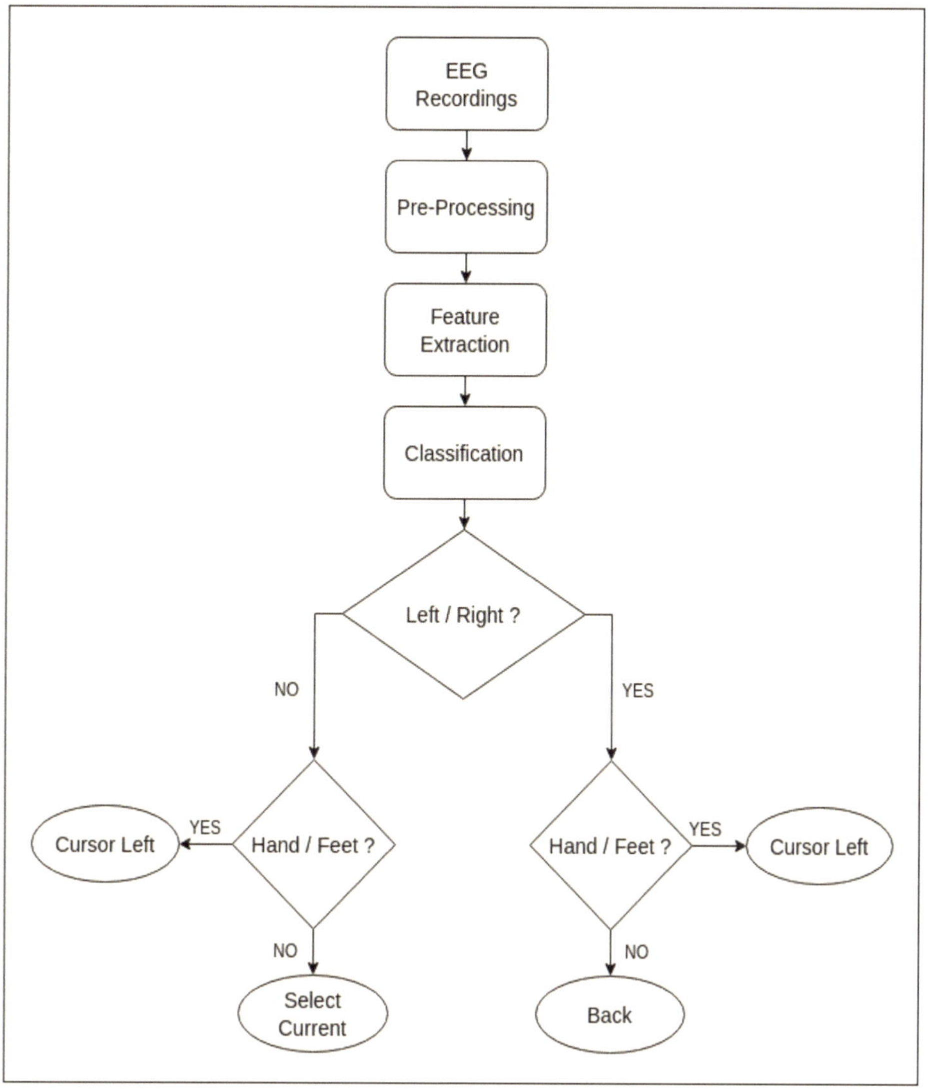

Figure 12.
Our BCI flowchart.

pre-processing and feature extraction. After cleaning, the data is labeled, and the data point is added to the dataset. After collecting all the data from all subjects, we fed this dataset for training. **Figure 11** shows flow of training module.

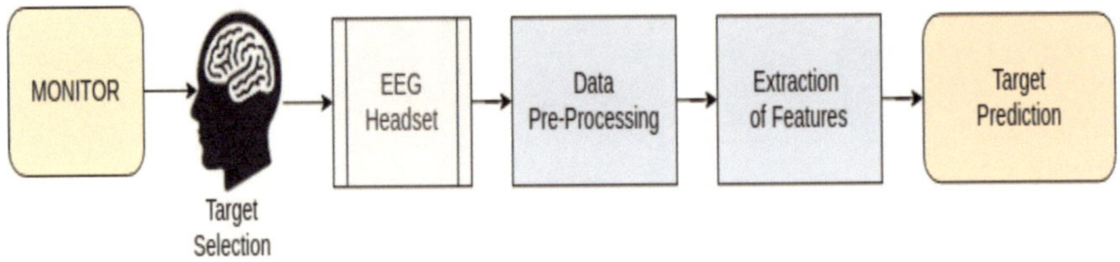

Figure 13.
Block diagram for BCI target prediction module.

We tried may classifiers like SVM with poly kernel but settled with Ensemble Subspace k-NN(k = 3) [26] with an accuracy of 91.4% on 5-fold cross validation (**Figure 12**). After the creation of the dataset and training, we run our model for real-time prediction. In the goal prediction pipeline, we continuously take raw data, store 1-sec data in a buffer, and do pre-processing and feature extraction. Feature-extracted unlabelled data is now fed to the same classifier that we use in the training stage for class(here goal) prediction. **Figure 13** shows the goal prediction module.

6. Results

We have taken data of eight healthy subjects for MI in two sessions as multi-session data collection improves overall real-world performance on new data. We take 100 readings of 5 seconds each, raw data for each subject in both the sessions, which on trimming 1 second and slicing the remaining in four, becomes 800 labeled data points per class. This dataset now go for training.

Accuracy of different classifiers at 5-fold cross validation is presented in **Table 1** below.

S.No.	Classifiers	Accuracy
1	k-NN (k = 3)	86.7%
2	Normal SVM	85.4%
3	SVM with poly kernel	88.2%
4	Ensemble Subspace kNN (k = 3)	91.4%

Table 1.
Performance comparison of different classifier models.

7. Discussion

The development of an advanced wheelchair is a significant breakthrough in enabling independent mobility for elderly and physically impaired individuals. However, it represents only a modest step towards the creation of an empowering and intelligent assistive technology.

There are certain limitations to our approach, such as the manual addition of key points within a known map by developers or admins, which the user can then use for BCI-based control. In contrast, a newly realized map can accept goal coordinates and orientation via manual touch entry. The main obstacle to achieving a higher level of freedom through BCI control is the intent prediction models associated with BCI. To overcome this, future research can focus on improving the granularity in motor imaging estimates to enable the selection of any point in all maps, both stored and realized, via BCI-based input.

8. Conclusion

In empirical investigations using Motor Imagery with healthy volunteers, we were able to achieve significant results within the confines of our laboratory's limited navigation space. Although the trials were successful in terms of goal acquisition, it may be necessary to retrain and redevelop the goal selection pipeline based on data obtained from motor-impaired individuals. Further research in this area could be expanded to include real-world settings such as hospitals and airports, in the hopes of establishing wider acceptance for this technology in the future. With these efforts, we envision a future where Motor Imagery becomes a widely recognized tool in the realm of rehabilitation for motor-impaired individuals.

While our wheelchair is well-suited for indoor use and provides excellent user convenience, additional improvements are necessary to ensure the safety of both the user and pedestrians in outdoor environments. A more efficient suspension and braking system can be developed to address this need. The current active obstacle avoidance system incorporates an 8000-sample 2D LiDAR sensor with a detection range of up to 6 m. For outdoor environments with larger navigable spaces and sparser point clouds, the obstacle detection range would have to be significantly increased. Our detection suite is supported by an Intel realsense depth camera, which has been tested and performs efficiently even in outdoor settings.

Our fundamental objective is to accentuate the gravity of the quandary we are attempting to assuage and accentuate the relatively facile approach by which an appropriate resolution can be reached to ensure autonomous mobility for geriatric and motor-challenged individuals. We ardently contend that targeted research and cutting-edge engineering solutions, focused on these innovative methodologies, will effectively ameliorate the quality of life for this particular demographic. Thus, we aim to meticulously scrutinize and develop these avant-garde techniques to revolutionize the way we facilitate independent movement for individuals with limited mobility. Through this, we aspire to make an indelible and significant contribution to the amelioration of society at large.

Author details

Ashok Kumar Chaudhary, Vinay Gupta, Kumar Gaurav, Tharun Kumar Reddy* and Laxmidhar Behera
Department of Electronics and Communication Engineering, Indian Institute of Technology Roorkee, Roorkee, India

*Address all correspondence to: tharun.reddy@ece.iitr.ac.in

References

[1] Iturrate I, Antelis JM, Kubler A, Minguez J. A noninvasive brain-actuated wheelchair based on a P300 neurophysiological protocol and automated navigation. IEEE Transactions on Robotics. 2009;**25**(3):614-627

[2] Rebsamen B, Guan C, Zhang H, Wang C, Teo C, Ang MH, et al. A brain controlled wheelchair to navigate in familiar environments. IEEE Transactions on Neural Systems and Rehabilitation Engineering. 2010;**18**(6): 590-598

[3] Müller ST, Celeste WC, Bastos-Filho TF, Sarcinelli-Filho M. Brain-computer interface based on visual evoked potentials to command autonomous robotic wheelchair. Journal of Medical and Biological Engineering. 2010;**30**(6): 407-415

[4] Millán JR, Renkens F, Mourino J, Gerstner W. Noninvasive brain-actuated control of a mobile robot by human EEG. IEEE Transactions on Biomedical Engineering. 2004;**51**(6):1026-1033

[5] Choi K. Control of a vehicle with EEG signals in real-time and system evaluation. European Journal of Applied Physiology. 2012;**112**:755-766

[6] Kim KT, Carlson T, Lee SW. Design of a robotic wheelchair with a motor imagery based brain-computer interface. In: 2013 International Winter Workshop on Brain-Computer Interface (BCI). Gangwon Province, South Korea: IEEE; 2013 Feb 18. pp. 46-48

[7] Duan J, Li Z, Yang C, Xu P. Shared control of a brain-actuated intelligent wheelchair. In: Proceeding of the 11th World Congress on Intelligent Control and Automation. Shenyang, China: IEEE; 2014 Jun 29. pp. 341-346

[8] Andronicus S, Harjanto NC, Widyotriatmo A. Heuristic steady state visual evoked potential based brain computer interface system for robotic wheelchair application. In: 2015 4th International Conference on Instrumentation, Communications, Information Technology, and Biomedical Engineering (ICICI-BME). Bandung, Indonesia: IEEE; 2015 Nov 2. pp. 94-97

[9] Zhang R, Li Y, Yan Y, Zhang H, Wu S, Yu T, et al. Control of a wheelchair in an indoor environment based on a brain–computer interface and automated navigation. IEEE Transactions on Neural Systems and Rehabilitation Engineering. 2015;**24**(1): 128-139

[10] Li Z, Zhao S, Duan J, Su CY, Yang C, Zhao X. Human cooperative wheelchair with brain–machine interaction based on shared control strategy. IEEE/ASME Transactions on Mechatronics. 2016; **22**(1):185-195

[11] Lahane P, Adavadkar SP, Tendulkar SV, Shah BV, Singhal S. Innovative approach to control wheelchair for disabled people using BCI. In: 2018 3rd International Conference for Convergence in Technology (I2CT). Pune, India: IEEE; 2018 Apr 6. pp. 1-5

[12] Xin L, Gao S, Tang J, Xu X. Design of a brain controlled wheelchair. In: 2018 IEEE 4th International Conference on Control Science and Systems Engineering (ICCSSE). Wuhan, China: IEEE; 2018 Aug 21. pp. 112-116

[13] Zgallai W, Brown JT, Ibrahim A, Mahmood F, Mohammad K, Khalfan M, et al. Deep learning AI application to an

EEG driven BCI smart wheelchair. In: 2019 Advances in Science and Engineering Technology International Conferences (ASET). Dubai, United Arab Emirates: IEEE; 2019 Mar 26. pp. 1-5

[14] Alkhatib R, Swaidan A, Marzouk J, Sabbah M, Berjaoui S, Diab MO. Smart autonomous wheelchair. In: 2019 3rd International Conference on Bio-Engineering for Smart Technologies (BioSMART). Paris, France: IEEE; 2019 Apr 24. pp. 1-5

[15] Nuo G, Wenwen Z, Shouyin L, Nuo G, Wenwen Z, Shouyin L. Asynchronous brain-computer interface intelligent wheelchair system based on alpha wave and SSVEP EEG signals. In: 2019 IEEE 4th International Conference on Signal and Image Processing (ICSIP). Wuxi, China: IEEE; 2019 Jul 19. pp. 611-616

[16] Dissanayake MG, Newman P, Clark S, Durrant-Whyte HF, Csorba M. A solution to the simultaneous localization and map building (SLAM) problem. IEEE Transactions on Robotics and Automation. 2001;**17**(3):229-241

[17] Grisetti G, Stachniss C, Burgard W. Improved techniques for grid mapping with rao-blackwellized particle filters. IEEE Transactions on Robotics. 2007; **23**(1):34-46

[18] Kohlbrecher S, Von Stryk O, Meyer J, Klingauf U. A flexible and scalable SLAM system with full 3D motion estimation. In: 2011 IEEE International Symposium on Safety, Security, and Rescue Robotics. Kyoto, Japan: IEEE; 2011 Nov 1. pp. 155-160

[19] Hess W, Kohler D, Rapp H, Andor D. Real-time loop closure in 2D LIDAR SLAM. In: 2016 IEEE International Conference on Robotics and Automation (ICRA). Stockholm, Sweden

[20] Mur-Artal R, Tardós JD. Orb-slam2: An open-source slam system for monocular, stereo, and rgb-d cameras. IEEE Transactions on Robotics. 2017; **33**(5):1255-1262

[21] Endres F, Hess J, Sturm J, Cremers D, Burgard W. 3-D mapping with an RGB-D camera. IEEE Transactions on Robotics. 2013;**30**(1): 177-187

[22] Pfurtscheller G, Da Silva FL. Event-related EEG/MEG synchronization and desynchronization: Basic principles. Clinical Neurophysiology. 1999;**110**(11): 1842-1857

[23] Zhang C, Lian Y, Wang G. ARDER: An automatic EEG artifacts detection and removal system. In: 2020 27th IEEE International Conference on Electronics, Circuits and Systems (ICECS). Glasgow, Scotland: IEEE; 2020 Nov 23. pp. 1-2

[24] Brigham EO, Morrow RE. The fast Fourier transform. IEEE Spectrum. 1967; **4**(12):63-70

[25] Alam MN, Ibrahimy MI, Motakabber SM. Feature extraction of EEG signal by power spectral density for motor imagery based BCI. In: 2021 8th International Conference on Computer and Communication Engineering (ICCCE). Kuala Lumpur, Malaysia: IEEE; 2021 Jun 22. pp. 234-237

[26] Bavkar S, Iyer B, Deosarkar S. Rapid screening of alcoholism: An EEG based optimal channel selection approach. IEEE Access. 2019;7:99670-99682

Chapter 3

Perspective Chapter: Classification of Grasping Gestures for Robotic Hand Prostheses Using Deep Neural Networks

Ruthber Rodríguez Serrezuela, Enrique Marañón Reyes, Roberto Sagaró Zamora and Alexander Alexeis Suarez Leon

Abstract

This research compares classification accuracy obtained with the classical classification techniques and the presented convolutional neural network for the recognition of hand gestures used in robotic prostheses for transradial amputees using surface electromyography (sEMG) signals. The first two classifiers are the most used in the literature: support vector machines (SVM) and artificial neural networks (ANN). A new convolutional neural network (CNN) architecture based on the AtzoriNet network is proposed to assess performance according to amputation-related variables. The results show that convolutional neural networks with a very simple architecture can produce accurate results comparable to the average classical classification methods and The performance it is compared with other CNN proposed by other authors. The performance of the CNN is evaluated with different metrics, providing good results compared to those proposed by other authors in the literature.

Keywords: electromyography, convolutional neural networks, support vector machine, artificial neural network, underactuated hand prosthesis

1. Introduction

Upper limb amputations are injuries that substantially limit a person's quality of life by drastically reducing the number of independent activities they perform in daily life (ADL). Current myoelectric prostheses are electronically controlled by the user's voluntary muscle contractions. A general scheme of how these and other devices that use biosignals work is presented in **Figure 1**. In this sense, the prostheses for amputees with higher performance follow this common pattern of development. There is a wide variety of very sophisticated myoelectric prostheses commercially available that use sEMG signals [1–5].

A relevant limitation in the development of pattern recognition methods for myoelectric control is that their tests are mainly performed offline. It is now established

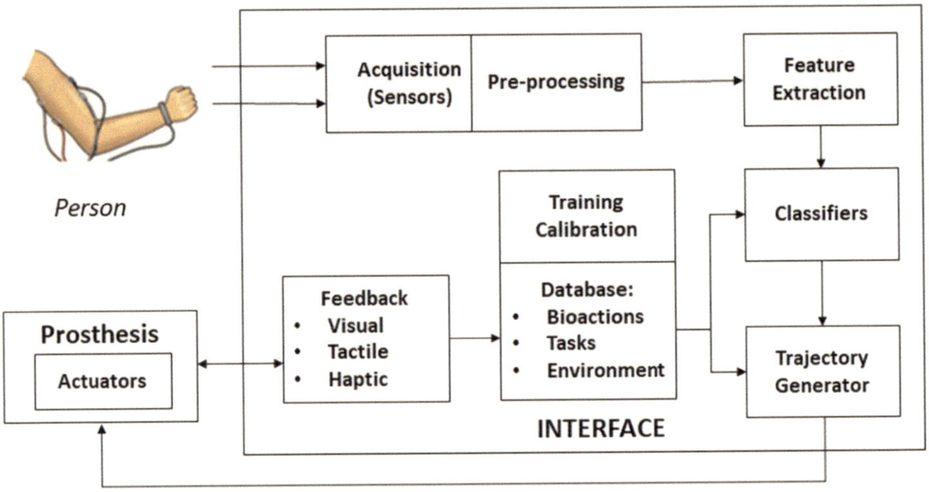

Figure 1.
Most common configuration of human–machine interaction [6].

that high offline precision, does not necessarily translate into accurate functional control of a physical prosthesis. In this sense, several recent studies have shown the discrepancy between "on and offline" in performance metrics [7–11]. However, very few studies have published validation results of pattern recognition in terms, and even fewer in clinical settings, relating the variability of the signal and the performance of the classifiers with those parameters related to amputation (disability index, length remaining limb, amputation time, phantom limb sensation, etc.) [12–15].

By other hand, the number of features extracted also depends on the number of EMG sensors and the feature extraction strategy for each sensor. Many investigations in have implemented alternatively, dimensionality reduction has been shown to be an effective feature projection (PC) method [16]. Among the most used methods are: principal component analysis (PCA) [17–19], linear-nonlinear PCA composite analysis, self-organizing feature maps [16] and supervised discretization together with PCA [20, 21].

Convolutional neural networks have been applied for myoelectric control with interest in inter-sessions/subjects and inter-session performance, in addition to many other applications in biomedical signal processing [22–24]. Some authors have commented on the advantages of these deep neural networks and their ability to assimilate the recognition of hand gestures corresponding to groups of sEMG signals. Although the results obtained come from a small number of investigations, their employment possibilities are promising [25–27].

However, most of the research has been carried out on healthy subjects. In recent decades, different authors [28–31] have shown that the variation of the signal over time in amputated patients is even greater than in healthy subjects. The EMG signal is weaker due to the amputation of certain muscle groups and as the amputation time elapses, the muscles become more atrophied and weak. There are also few databases of amputees, a situation that constitutes a significant obstacle for these researches. and for the gestures recognition at the international level [29, 30]. Additionally, amputees' performance was found to be proportional to residual limb size, indicating that an anthropomorphic model might be beneficial [28–31]. The previous findings motivated the study on the variance of results between amputee patients and fit populations under disturbances of dynamic factors such as the length of the remaining limb, age, level of atrophy, among others. That is why the results obtained in amputated patients are far from those presented.

2. Materials and methods

2.1 Databases subjects

The review of these databases allows knowing the characteristics of the population involved and the signal capture protocols. The literature review showed that there are few databases with sEMG data collection in a significant number of patients, with subjects without known prior deficiencies and whose data are heterogeneous, so the most used is the NINAPRO database [32–34], which contains the electromyography recording using the system of 8 sEMG sensors Thalmic Labs - MYO. The data in this repository is free to use and is intended for use in developing hand gesture movement classifiers [22]. The NINAPRO database in its DB3 section establishes the parameters with which the sEMG data of 11 subjects with transradial amputation were recorded [35].

In the DB3 dataset, as explained above, the transradial amputee wears two Myo cuffs side by side. The superior MYO cuff is placed closest to the elbow with the first electrode at the radio-humeral joint, following the configuration of the NINAPRO electrode. The lower MYO cuff is placed just below the first one, closer to the amputation region (**Table 1**).

In order to build our own database, the subjects invited to participate in this stage are amputated subjects, without neurological deficiencies. Invited subjects followed the population parameters used in the NINAPRO Database [36, 37]. Ten male and

Patient	Hand	Laterality	Age	Remained Forearm (%)	Years since amputation	Amputation cause	DASH Score	Time wearing prostheses (years)
1	Right	right handed	32	50	13	Traumatic injury	1.67	13
2	Left	right handed	35	70	6	Traumatic injury	15.18	6
3	Right	right handed	50	30	5	Traumatic injury	22.50	8
4	Right	right handed	34	40	1	Traumatic injury	86.67	0
5	Left	left handed	67	90	1	Traumatic injury	11.67	0
6	Left	right handed	32	40	13	Traumatic injury	37.50	0
7	Right	right handed	35	0	7	Traumatic injury	31.67	0
8	Right	right handed	33	50	5	Traumatic injury	33.33	0
9	Right	right handed	44	90	14	Traumatic injury	3.33	0
10	Right	right handed	59	50	2	Traumatic injury	11.67	0
11	Right	right handed	45	90	5	Cancer	12.50	0

Table 1.
Clinical characteristics of subjects with amputation. NinaPro DB3.

female amputees ranging in age from 24 to 65 years participated in the experiments. The procedures were performed in accordance with the Declaration of Helsinki and were approved by the ethics committee of the Universidad del Tolima (approval number: N – 20,160,021). All subjects participated voluntarily, providing their written informed consent before the experimental procedures. Any amputee who has experience in the use of hand prostheses will be included in the study, registering in advance their experience in the use of passive or myoelectric prostheses.

Aspects and demographic data to be recorded: For each subject, age, sex, education level, related to the amputation: dominant hand, amputated side, year of amputation, cause, type of prosthesis if used or has been used, and level of amputation.

Inclusion criteria: Adults in an age range of 20–65 years, no history of neurological and/or psychiatric diseases, voluntary participation in the study and acceptance of the medical staff. Only the transradial level of amputation will be considered, amputations above the elbow or beyond the wrist will not be admitted to the study. Any non-compliance with these parameters becomes criteria for exclusion from the study. **Table 2** shows the characteristics of the amputee patients who participated in the trials.

2.2 Sensor EMG MYO armband

Data was recorded using the commercial MYO armband (MYO). MYO is a portable EMG sensor developed by Thalmic Lab and has eight dry electrode channels with a sampling rate of 200 Hz. It is a low cost, consumer grade device with a nine inertial measurement unit (IMU) [22], that connects wirelessly with the computer via Bluetooth. It is a non-invasive device, easier to use compared to conventional electrodes [38, 39]. Despite the low sampling frequency, its performance has been shown to be similar to that of full-band EMG recordings using conventional electrodes [22, 40], and the technology has been used in many studies [29, 35, 38] (**Figure 2**).

sEMG recording: Prior to carrying out the tests, the patients will be instructed on the experimental procedure and as a first step. The sensor operation will be calibrated for both limbs. To make the records in each gesture, the subjects will be seated comfortably in front of the computer with both limbs with their elbows flexed at 90

Patient	Age (years)	Gender	Remained Forearm Length (Below elbow)	Amputated hand	Years since amputation	DASH Score
P01	36	M	10 cm	Dominant hand	1	45
P02	51	M	Wrist Disarticulation	Dominant hand	30	19
P03	62	M	Wrist Disarticulation	Dominant hand	36	39.16
P04	26	M	10 cm	Dominant hand	12	20
P05	60	M	Wrist Disarticulation	Not dominant Hand	41	26.6
P06	55	M	Wrist Disarticulation	Dominant hand	5	16.66
P07	28	M	10 cm	Dominant hand	9	24.16
P08	48	F	Wrist Disarticulation	Dominant hand	22	20.83
P09	65	M	Wrist Disarticulation	Dominant hand	29	42.5
P10	35	M	10 cm	Dominant hand	2	47.5

Table 2.
Clinical characteristics of subjects with amputation.

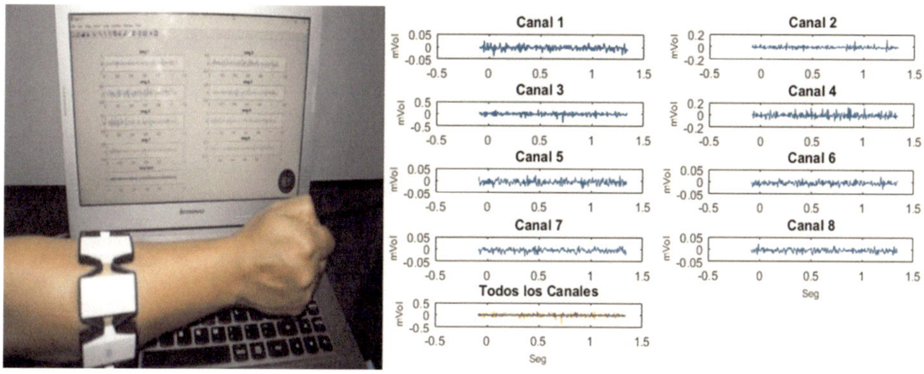

Figure 2.
*Signals acquisition through the application developed in Matlab 2020b. Author. Six movements were identified in MYO sensor to achieve grip improvement: power grip (AP), palm inward (PI), palm outward (PO), open hand (MO), pincer grip (AT) and rest (RE) (**Figure 3**).*

degrees and will be instructed to perform the gestures that are reflected on the monitor, in the case of amputated patients with the contralateral limb and with the amputated limb (**Figures 4** and 5).

The graphic interface will provide the patient with the times for performing the tests and the state of rest (**Figure 5**). Amputee recordings were performed in repeated sessions for 1 week.

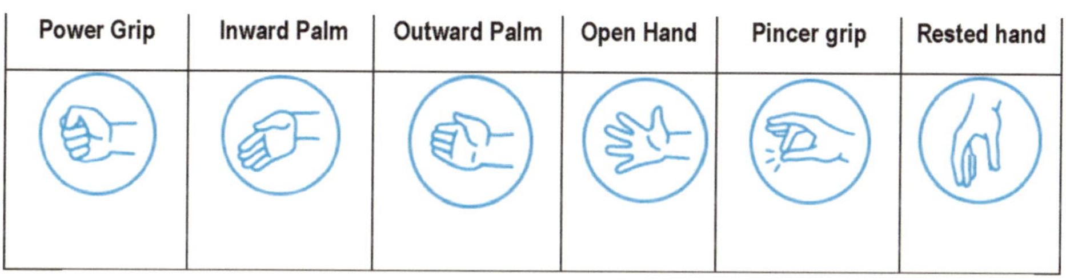

Figure 3.
Gestures to identify with the MYO device.

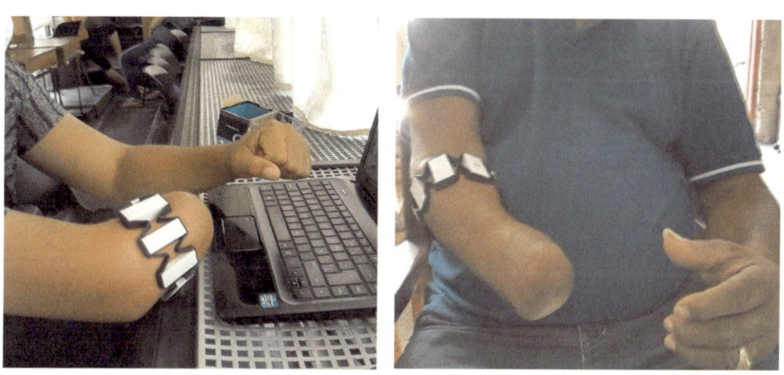

Figure 4.
(a) Amputee patient in front of the computer with a graphic signal to perform the movements. (b) MYO device arrangement.

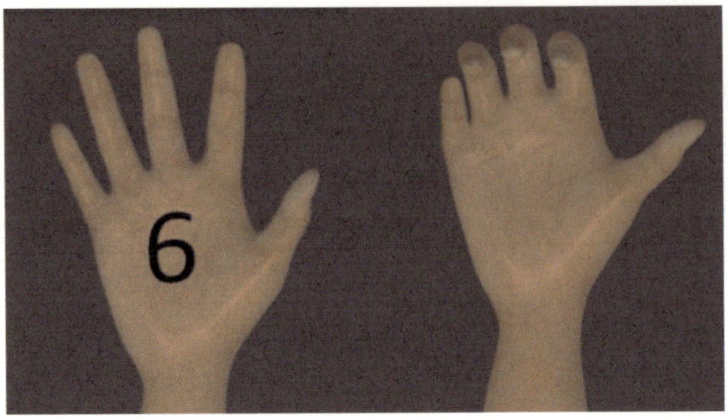

Figure 5.
User interface that indicates the imaginary movement to be performed and includes completion and rest time.

The procedure carried out to capture the myoelectric signals is as follows: for each grip or gesture of the hand, 200 samples are taken during an interval of 30 seconds. Transitions are made between each of the six proposed gestures for spaces of 1 minute as recommended in [41]. sEMG signals were captured during several sessions and on different days of the week. The data of these myoelectric signals are stored in dataset for later offline processing.

2.3 Signal pre-processing

The segmentation and overlay methods used in this work improved the training efficiency by increasing the number of training samples based on recent work such as [17, 20, 42].

2.4 Feature extraction

Each captured sEMG signal is subdivided into 200 ms windows. The signal captured by the MYO was obtained at a frequency of 200hz [11, 21]. In order to be analyzed, it is divided every 200 ms, leaving a total of 40 data in each sub-window. Each sub-window has a 50% overlap with the immediately previous window, which allows increasing the number of samples and thus expanding the database obtained. The extraction described here is applied to each of the MYO channels. These data obtained for each of the channels [19], are concatenated horizontally, thus allowing a database to be obtained with 10 data features for each channel and a column with the information on the grasping gestures that are performed.

Different kinds of features extracted are used by different researchers, such as mean absolute value (MAV), root mean square (RMS), autoregression coefficients (AC), variance (VAR), standard deviation (SD), crossover by zero (CC), the length of the waveform (LO), Amplitude of Wilson (AW), slope of mean absolute value (PVAM). Features in the time domain were treated in [42]. These extracted features are used in SVM and in ANN and the raw signals for the CNN classifier.

2.5 Classifiers

Artificial neural networks Artificial neural networks (ANN) are a nonlinear classifier that simulates brain information processing through a series of weighted nodes, called neurons. Neurons are organized in layers and interconnect with each other to create a network. ANNs use a nonlinear function of a linear combination of the inputs, where the coefficients of the linear combination are adaptive parameters. The basic model of an ANN will be described as a series of functional transformations. First, M linear combinations of the input variables are constructed $x_1, x_2, \dots, x_D$ in a way:

$$a_j = \sum_{i=1}^{D} w_{ji}^{(1)} + w_{j0}^{(1)} \tag{1}$$

where $j = 1, \dots, M$, y the superscript (1) indicates that the corresponding parameters correspond to the first layer of the network. Parameters $w_{ji}^{(1)}$ are weights and the parameters $w_{j0}^{(1)}$ are polarization constants. The magnitude a_j is named activation, and each activation is transformed using a nonlinear and differentiable activation function [22].

In ANN classifier (**Table 3**), the chosen hyper parameters are highlighted and the variations that were proposed for each of them. Among the hyper parameters that were varied are the weight optimizer, the activation method, the maximum number of iterations and the type of learning.

2.6 Support vector machine

Support vector machines are a very powerful method used to solve classification problems, it is highly efficient on multidimensional data sets. It consists of defining a hyperplane or decision limit that separates the samples into two groups, where those above the decision limit are classified as positive and those below it, are classified as negative. The fundamental objective is to maximize the margin M or distance between the neighboring samples called support vectors and the separating hyperplane (**Figure 6**) [43].

For the SVM classifier, the following kernels were used: linear, polynomial, Gaussian and sigmoid as shown in **Table 4**. In the linear and Gaussian kernels, the Gamma parameter was used with a value of 0.5. In the polynomial and sigmoid nuclei, the Degree parameter was used: between: 0.5 and 3, respectively. For all nuclei, the regularization constant C was used with values of 0.1, 1, 10 and 100 as recommended in the literature [44, 45].

2.7 Convolutional neural networks

An CNN is a deep learning algorithm able to collecting an input matrix of size M X N and attributing weights and biases in parallel under the constraints of a predictive

Parameters	Options
Optimizer	SDG, **ADAM**
Regularization	**0.0001**–0.00001
Activation Function	Tanh, **ReLu**, Logístico
maximum numbers of iterations	10, 30, 50, **100**
Hidden layers	3, **4**, 5 y 6
Learning rate	Constant, Adaptive

Table 3.
Hyper-parameters. ANN.

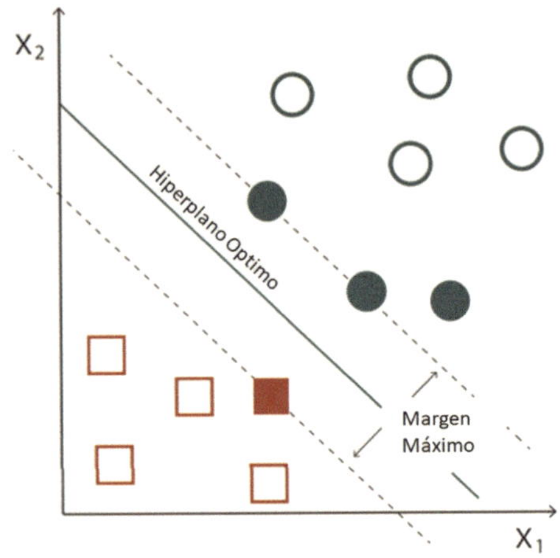

Figure 6.
Optimal separation hyperplane, for linearly separable classes.

Parameters	Options
Kernel	Lineal, Gaussian, Polynomial, Sigmoid
Gamma	0.5
Degree	0.5–3
C	0.1, 1, 10 y 100

Table 4.
SVM classifier parameters.

problem [46], resulting in specific features. A convolutional layer performs a dot product between two arrays, where one array is the set of learnable parameters and the other is known as the kernel, producing an activation map, as shown below:

$$G[m,n] = (f,h)[m,n] = \sum_{j}^{M}\sum_{k}^{N} h[j,k], f[m-j, n-k], \tag{2}$$

Where:
The input matrix is f and the kernel is denoted as h.
m is number of rows in the input matrix
n is the number of columns in the input matrix
j is the index for the offset in the rows of the input
k is the index for the offset in the columns of the input (**Figure 7**)

Table 5 shows the parameters used in the CNN that were selected for the experiment carried out. Highlighted in the text are the hyper parameters: the batch size and the number of epochs that yielded the best results. The batch sizes were worked with values of 64, 128 and 256 respectively. The CNN architecture has four hidden convolutional layers and one output layer (**Figure 8**).

The first two hidden layers consist of 32 filters of size 8x1 and 3 × 3. The third consists of 64 filters of size 5 × 5. The fourth layer contains 64 filters of size 5 × 1, while the last one is a convolutional layer with six possible outputs with 1 × 1 filters,

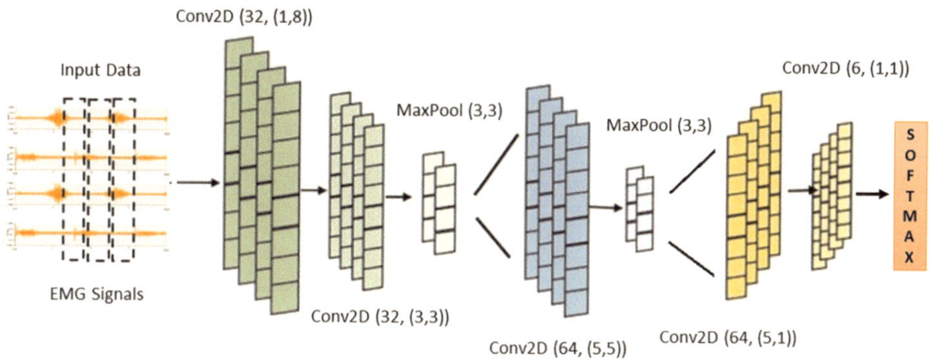

Figure 7.
Architecture of the convolutional neural network used.

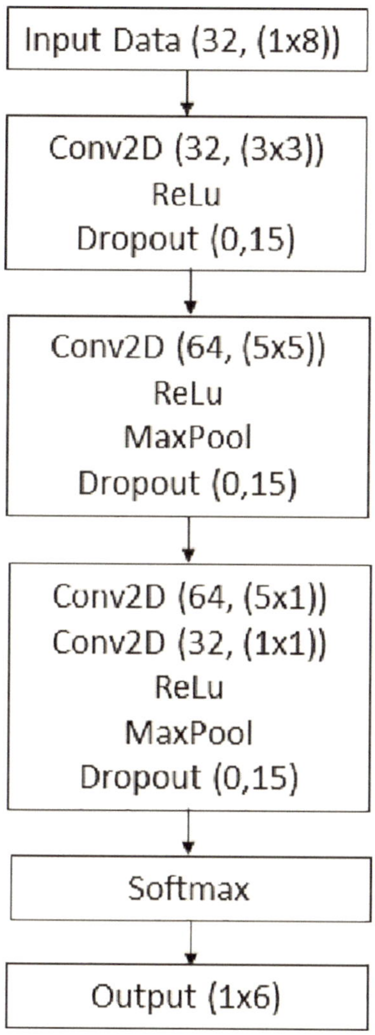

Figure 8.
Proposed model of CNN.

corresponding to the six gestures to classify. Followed by Nonlinear Rectification Units (ReLU) and Dropout layer with a probability of 0.15 to put to zero the output of a hidden unit. Also, a subsampling layer performs maximum pooling in a 3×3 window after removal of the second and third layers. Finally, the last convolutional layer is followed by a Softmax activation function (**Table 5**).

2.8 Metrics

The confusion matrix is used to calculate many common classification metrics. The diagonal represents correct predictions and the other positions of the matrix indicate incorrect predictions. If the sample is positive and is classified as positive, i.e. correctly classified positive sample, and it is considered as a true positive (TP); if it is classified as negative, it is considered a false negative (FN). If the sample is negative and is classified as negative, it is considered true negative (TN); If the sample is negative and is classified as negative, it is considered true negative (TN); if it is classified as positive, it is counted as a false positive (FP), false alarm. The most common metrics are sensitivity (Se), specificity (Sp), which indicate the ability of the CNN to identify hand gestures. Accuracy (ACC) is used to assess overall detection performance and Precision (Pr) is used to measure model quality in posture classification tasks. Likewise, the F1 score (F1) is a measure of the precision of a test, it is the average precision and sensitivity of the classification. It has a maximum score of 1 (perfect accuracy and sensitivity) and a minimum of 0. Overall, it is a measure of the accuracy and robustness of your model. This metric gives information about the quantity that the model is able to identify. In this work, six commonly used evaluation metrics were used: accuracy, precision, sensitivity, specificity and F1 to evaluate the performance of CNN:

$$Accuracy\,(ACC) = \frac{TP + TN}{TP + TN + FP + FN} \tag{3}$$

$$Precision\,(\mathrm{Pr}) = \frac{TP}{TP + FP} \tag{4}$$

$$\mathrm{Sensitivity}\,(Se) = \frac{TP}{TP + FN} \tag{5}$$

$$\mathrm{Specificity}\,(Sp) = \frac{TN}{TN + FP} \tag{6}$$

$$F1score\,(F1) = \frac{2^{*}TP}{TP + FP + FN} \tag{7}$$

Parameters	Selected choice
Lost function	Categorical cross entropy
Optimizer	ADAM (lr = 0.001)
Batch size	64, 128, 256
Epoch	100, 400, 500, 1000
Validation rate	30%

Table 5.
Hyper parameters tuning.

3. Results

The analysis in amputated patients is preceded by the great variability of the sEMG signal. **Figure 9** shows this behavior of the sEMG signals for the power grab gesture in amputated subjects. It is observed that the data are very scattered and do not have the same symmetry, either the same mean, either standard deviation between each one of patients.

3.1 ANN classifier

The results obtained with this classifier are shown in **Table 6** in the specific patients. The own database has been used for the test data set with the following configuration: Optimizer: ADAM, activation function: ReLu, L2 regularization with a constant of 0.0001, a constant learning rate of 0.001, with three hidden layers and all layers with eight neurons.

On both data sets, that is, the test and trial data set, the ANN classifier showed an increase in the accuracy metric, especially when increasing the number of training

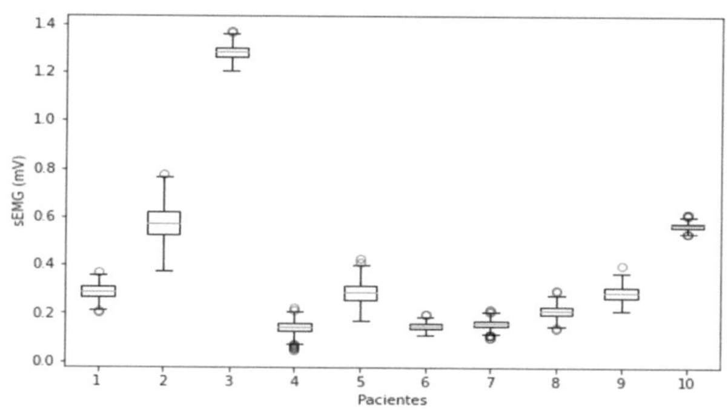

Figure 9.
Box and Whisker Plot of the sEMG for the power grab gesture in amputees.

	ANN	
Patients	**Accuracy**	**Lost**
P01	85.71	0.1000
P02	80.22	0.1700
P03	56.04	0.0280
P04	82.00	0.8600
P05	76.92	0,4700
P06	79.12	0.1886
P07	85.71	0.1200
P08	72.53	0.1920
P09	67.03	0.2111
P10	81.32	0.1239

Table 6.
Accuracy results with the ANN classifier on the test data set at 100 epochs.

epochs, passing on average from 76.66% to 100. periods with a minimum of 56.04% and a maximum of 85.77%, respectively. The average accuracy was consistent with the results shown by other authors for this classifier [35, 44, 45, 47].

In **Table 6**, superior results can also be seen on subjects P01, P02, P04, P05, P07 and P10, which is outstanding, since their accuracy was above 80% with this classifier with a standard deviation of 9.23. This low standard deviation indicates that most of the accuracies obtained tend to be clustered close to their mean.

3.2 SVM classifier

As mentioned above, different kernels were used: linear, polynomial, Gaussian and sigmoid. Likewise, we worked on the regularization constant C with values of 0.1, 1, 10 and 100. On both data sets with the SVM classifier using the RBF and Sigmoid kernels, the best results were obtained when evaluating the accuracy metric, obtaining results up to 80%. in this metric. **Table 7** shows the results obtained with all the kernels.

3.3 CNN Classifier

Table 8 shows the comparative results of the CNN classifier evaluated in different patients using regularization techniques such as: early stopping, dropout and batch normalization. Which is a technique that tries to apply certain rules to know when it is time to stop training, so that there is no overfitting to the input data, nor under fitting. The average time required to train each convolutional neural network was 1 hour and 25 minutes. The average time needed to test the network was 15.2 s using a GPU NVIDIA Titan-V, 12 GB RAM HBM2 y 640 Tensor Cores.

Table 9 shows a summary of the different classifiers by means of the accuracy metric in the different patients. In patients P01 and P02, the best classifier is ANN. Although the accuracy shown by the CNN is acceptable. These patients have a DASH index of 45 and 19, respectively. They also present different levels of amputation: one at 10 cm from the elbow and the other at the level of the wrist. Likewise, they present amputation times of 1 and 30 years, correspondingly. These clinical factors affect the performance of the classifiers.

4. Discussion

From the results obtained, the following points are analyzed. **Figure 10** presents the confusion matrix of each of the classifiers (SVM, ANN and CNN) corresponding to the patient (P03 0 P09), It is observed that both the SVM and ANN show a low number of hits in the different gestures in this patient whose cause of amputation is due to congenital factors that further affect the variability of the signal. Even though only one case is analyzed in this work, this type of behavior has been reported by other research works in patients of this type.

Once again, the performance of the SVM and the ANN is significantly affected. The results of the present work show a significant accuracy rate for the classification of various classes of amputated subjects in comparison with other studies carried out (**Table 10**).

DOI: http://dx.doi.org/10.5772/intechopen.107344

		C = 0.1				C = 1				C = 10				C = 100			
Patients	**Kernel**	**Accuracy**	**F1 score**	**Sensitivity**	**Precision**	**Accuracy**	**F1 score**	**Sensitivity**	**Precision**	**Accuracy**	**F1 score**	**Sensitivity**	**Precision**	**Accuracy**	**F1 score**	**Sensitivity**	**Precision**
P01	**Polynomial**	47.5	41.25	51.39	48.23	47.5	51.62	58.06	51.64	25.00	23.94	28.94	29.12	25.00	23.94	28.94	69.12
	Linear	**62.5**	**59.82**	**64.63**	**60.89**	**62.5**	**61.17**	**65.32**	**62.74**	25.00	17.29	30.00	25.00	25.00	20.21	29.86	27.50
	RBF	12.5	2.78	16.67	2.08	22.5	19.36	27.08	52.38	**65.00**	**62.66**	**67.55**	**66.44**	**65.00**	**59.82**	**64.63**	**60.89**
	Sigmoid	12.5	2.78	16.67	12.5	12.5	18.21	33.33	27.5	60.00	59.07	63.24	61.55	60.00	62.39	65.09	64.60
P02	**Polynomial**	20.00	9.20	25	7.73	20	30.97	32.16	30.86	20.00	15.87	23.21	43.98	20.00	15.87	23.21	43.98
	Linear	**52.5**	**52.78**	**52.46**	**59.76**	**52.5**	**49.77**	**51.03**	**57.13**	2.50	1.74	3.33	2.50	2.50	2.15	6.11	5.00
	RBF	12.5	2.78	16.67	2.08	10	2.7	13.33	2.08	**42.50**	**35.85**	**47.16**	**38.57**	**42.50**	**36.37**	**47.16**	**47.62**
	Sigmoid	12.5	2.78	16.67	12.5	12.5	2.78	16.67	12.5	35.00	41.42	43.27	42.26	35.00	30.44	38.27	31.15
P03	**Polynomial**	32.50	24.25	36.30	33.65	32.5	37.23	42.22	34.72	20.00	13.21	25.00	19.12	20.00	13.21	25.00	19.12
	Linear	**55**	**49.47**	**55.6**	**46.89**	**55**	**49.47**	**55.6**	**46.89**	5.00	3.13	6.67	5.00	5.00	10.3	10.2	11.5
	RBF	12.50	2.78	16.67	2.08	20	13.21	25	19.12	**55.00**	**50.33**	**55.60**	**48.28**	55.00	47.09	49.49	44.44
	Sigmoid	12.50	2.78	16.67	12.5	12.5	2.56	13.33	10	50.00	44.78	49.63	45.69	50.00	46.61	52.27	43.22
P4	**Polynomial**	47.5	42.18	47.22	39.58	47.5	64.15	65.65	65.93	20.00	15.99	24.31	52.25	20.00	15.99	24.31	52.25
	Linear	**65**	**64.36**	**64.07**	**69.83**	**65**	**64.36**	**64.07**	**68.83**	2.50	3.91	2.78	2.50	2.50	2.80	2.70	2.30
	RBF	12.5	2.79	16.67	2.08	17.5	11.48	22.22	35.53	65.00	64.15	64.07	68.75	**67.50**	**67.24**	**68.10**	**70.24**
	Sigmoid	12.5	2.78	16.67	12.5	12.5	2.78	16.67	12.5	**67.50**	**66.82**	**67.41**	**71.53**	65.00	68.25	66.85	70.83
P05	**Polynomial**	17.5	7.06	22.22	5.65	17.5	27.37	38.33	28.84	20.00	14.75	25.00	33.16	20.00	14.75	25.00	33.16
	Linear	**52.5**	**50.76**	**56.62**	**53.89**	**52.5**	**53.28**	**59.95**	**56.83**	12.50	3.29	16.67	12.50	12.50	3.97	5.56	5.00
	RBF	12.5	2.78	16.67	2.08	20	14.66	25	33.08	**55.00**	**53.74**	**59.95**	**57.45**	**55.00**	**50.80**	**57.87**	**54.99**
	Sigmoid	12.5	2.78	16.67	12.5	12.5	2.78	16.67	12.5	40.00	37.70	44.35	40.08	40.00	51.51	56.62	55.40
P06	**Polynomial**	32.5	25.44	36.3	36.31	32.5	41.29	50.37	40.1	25.00	19.41	30.56	35.71	25.00	19.41	30.56	35.71
	Linear	**65**	**65.85**	**65.32**	**74.63**	**65**	**65.85**	**65.32**	**74.63**	2.50	4.17	3.33	2.50	**2.50**	**3.10**	**4.20**	**3.70**
	RBF	12.5	2.78	16.67	2.08	17.5	10.41	22.22	18.86	57.50	51.28	60.56	49.81	57.50	57.21	61.16	62.82
	Sigmoid	12.5	2.78	16.67	12.5	12.5	2.78	16.67	12.5	**65.00**	**65.85**	**65.32**	**74.63**	**65.00**	**68.26**	**68.10**	**75.32**

		C = 0.1				C = 1				C = 10				C = 100			
P07	Polynomial	75	70.18	71.51	63.07	75	79.18	76.51	79.54	27.50	31.06	29.64	54.44	27.50	31.06	29.64	54.44
	Linear	**80**	**79.76**	**77.76**	**79.4**	**80**	**79.76**	**77.76**	**79.4**	15.00	8.94	15.48	15.00	15.00	9.37	15.48	15.00
	RBF	12.5	2.78	16.67	2.08	17.5	14.5	20.87	35.19	**80.00**	**65.77**	**64.56**	**66.05**	**80.00**	**66.04**	**63.87**	**69.44**
	Sigmoid	12.5	2.78	16.67	12.5	12.5	19.36	30.95	27.5	62.50	63.13	60.40	68.06	62.50	63.35	60.54	63.33
P08	Polynomial	45	35.28	51.67	34.47	45	44.13	55.37	52.15	25.00	20.71	30.56	35.78	25.00	20.71	30.56	35.78
	Linear	**45**	**46.43**	**46.71**	**50.89**	**45**	**49.32**	**49.49**	**55.51**	17.50	10.06	21.67	17.50	17.50	9.47	18.33	15.00
	RBF	12.50	2.78	16.67	2.08	17.50	10.48	22.22	18.92	**57.50**	**52.28**	**59.44**	**50.67**	**57.50**	**49.79**	**58.15**	**48.96**
	Sigmoid	12.50	2.78	16.67	12.50	12.50	12.98	25	20	47.50	45.76	49.49	49.74	47.50	40.89	47.41	39.98
P9	Polynomial	40	38.25	43.33	48.57	40	41.75	42.78	44.7	17.50	11.19	22.22	25.00	17.50	11.19	22.22	25.00
	Linear	**45**	**42.38**	**45**	**40.37**	**45**	**44**	**45**	**43.94**	10.01	9.27	7.35	8.12	**12.0**	**9.27**	**8.48**	**12.86**
	RBF	12.5	2.78	16.67	2.08	15	6.81	19.44	8.24	45.00	42.72	45.00	40.42	45.00	47.82	53.33	46.52
	Sigmoid	12.5	2.78	16.67	12.50	12.50	3.01	10.00	7.50	**47.50**	**43.71**	**47.22**	**41.94**	**47.50**	**38.56**	**38.89**	**36.55**
P10	Polynomial	35	26.76	40.56	30.11	35	77.21	77.59	78.02	22.50	19.57	26.16	52.38	22.50	19.57	26.16	52.38
	Linear	**72.5**	**72.9**	**71.25**	**72.02**	**72.5**	**73.07**	**71.25**	**72.32**	12.50	7.86	14.44	12.50	**12.50**	**9.79**	**14.44**	**12.50**
	RBF	12.5	2.78	16.67	2.08	20	15.05	24.07	35.65	**75.00**	**75.21**	**74.03**	**74.40**	**75.00**	**70.70**	**69.40**	**70.06**
	Sigmoid	12.5	2.78	16.67	12.5	12.5	12.44	21.67	17.5	75.00	75.05	73.10	73.21	75.00	69.68	69.40	70.00

Table 7.
Results of the metrics with the SVM classifier in the specific patients.

	CNN									
	EPOCHS									
	100	200	300	400	500	600	700	800	900	1000
Patients	Accuracy	Accuracy	Accuracy	Accuracy	Accuracy	Accuracy	Accuracy	Accuracy	Accuracy	Accuracy
P01	61.53	64.87	67.38	69.26	71.61	73.86	76.08	78.11	80.26	80.51
P02	54.15	58.23	60.16	60.80	63.65	67.76	69.57	72.98	76.13	76.52
P03	55.32	60.41	60.94	62.84	64.54	68.71	70.97	73.64	75.95	76.34
P04	67.28	70.46	72.20	72.91	74.09	76.24	79.39	81.54	83.12	83.69
P05	60.07	62.99	65.62	67.08	68.38	71.64	73.91	76.28	77.95	78.58
P06	55.53	59.51	62.64	62.51	67.37	69.61	71.89	74.32	76.54	78.85
P07	70.44	73.25	74.61	76.54	77.12	78.88	80.65	82.51	84.62	86.63
P08	56.51	60.80	63.30	64.81	65.87	67.97	71.11	74.25	77.39	80.53
P09	56.72	63.53	63.20	66.44	69.83	71.37	73.86	76.38	78.54	81.20
P10	62.04	67.71	70.01	71.39	73.27	74.85	76.58	78.31	80.14	81.77

Table 8.
Comparative results of the CNN classifier of the different patients at different epochs.

Patients	SVM (Accuracy %)	ANN (Accuracy %)	CNN (Accuracy %)
P01	**65.00**	**85.71**	80.51
P02	**52.20**	**80.22**	76.52
P03	55.00	56.04	**76.34**
P04	67.50	82.00	**83.69**
P05	**55.00**	76.92	**78.58**
P06	65.00	**79.12**	78.85
P07	80.00	**90.11**	86.63
P08	**57.50**	**84.62**	80.53
P09	**47.50**	67.03	**81.20**
P10	75.00	81.32	**81.77**
Average	61.97	78.30	**80.46**

Table 9.
Accuracy metric comparison between all classifiers. The values in bold represent the highest accuracy values recorded by each patient in the classifiers.

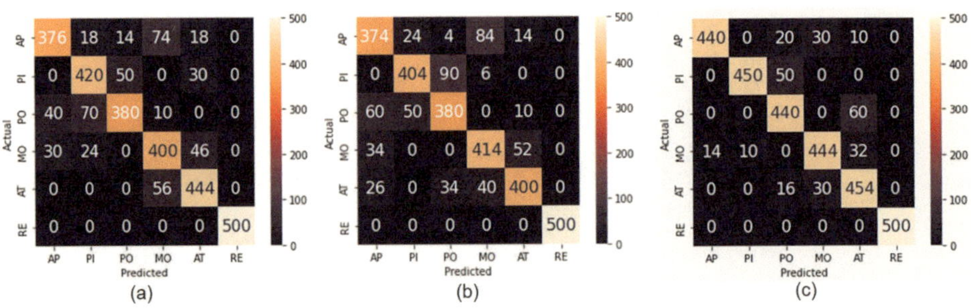

Figure 10.
Confusion matrices of the different classifiers for the patient P09 (a) SVM (b) ANN (c) CNN using the metric of the accuracy of the data corresponding to the training (session two).

Methods	Accuracy	Authors
AtzoritNet, CNN classifier, healthy subjects and amputees	70.48±1.52%	Tsinganos et al., 2018 [39]
Time domain, regressive models, Bayesian Network, ANN, CNN, AD, SVM, healthy subjects	80.4±2.6%	Ramirez-Martinez, et al., 2019 [40]
CNN, multiclass classifier, input TWC, amputees subjects	68.98±29.33%	Cote-Allard et al., 2019 [22]
WeiNet, CNN classifier, NinaPro dataset, healthy subjects	99.1%	Wei et al., 2020 [27]
AD Classifier, ACP, MD, Multiclass Classifier, amputees patients	77.3±17.5%	Rabin Neta et al., 2020 [38]
SVM classifier, LDA and TWD, healthy subjects	94.73%	Lin Chen et al., 2020 [48]
CNN classifier, TWD, healthy subjects	98.82%	Tsinganos et al., 2020 [49]
Multiclass classifiers, SVM, ANN, CNC, amputees patients	80.46%	This research

Table 10.
Studies conducted using CNN as an EMG-based hand prosthesis movement classifier in healthy subjects and amputated subjects.

5. Conclusions

Over the past decade, deep learning and convolutional neural networks have revolutionized several fields of machine learning, including speech recognition and computer vision. Therefore, its use is promising for obtaining better classification indexes of the sEMG signals if the great variation of these is considered in accordance with clinical variables of the amputation, all of which would contribute to closing the gap between the prosthesis market (which requires fast and robust control methods) and the results of recent scientific research in disability support technologies.

The protocol for obtaining sEMG measurements in amputated patients was applied, as well as the extraction and classification of the signal, all of which is consistent with the proposal for the integrated design of the prosthesis. A database of 10 amputated patients according to the 6 defined hand gestures was constructed. The data is publicly available in the repository of the Huila-Corhuila University Corporation (CORHUILA).

The classification accuracy obtained with CNN using the proposed architecture is 80.46%, but the most significant thing is its ability to obtain a higher performance in the classification between subjects in relation to parameters such as length of the remaining limb, years of amputation or disability index, compared with the results obtained by conventional classifiers such as the support vector machine and artificial neural networks.

Author details

Ruthber Rodríguez Serrezuela[1], Enrique Marañón Reyes[2], Roberto Sagaró Zamora[3]* and Alexander Alexeis Suarez Leon[4]

1 University Corporation of Huila, Colombia

2 Centro de Estudios de Neurociencias y Procesamiento de Imágenes y Señales, Universidad de Oriente, Santiago de Cuba, Cuba

3 Departamento de Mecánica y Diseño (MyD), Tribology Group, Universidad de Oriente, Santiago de Cuba, Cuba

4 Biomedical Engineering Department, Universidad de Oriente, Santiago de Cuba, Cuba

*Address all correspondence to: sagaro@uo.edu.cu

References

[1] Fukaya N, Asfour T, Dillmann R, Toyama S. Development of a five-finger dexterous hand without feedback control: The TUAT/Karlsruhe humanoid hand. In: Intelligent Robots and Systems (IROS), 2013 IEEE/RSJ International Conference on. IEEE. November 2013. pp. 4533-4540

[2] Diftler MA, Mehling JS, Abdallah ME, Radford NA, Bridgwater LB, Sanders AM. et al. Robonaut 2-the first humanoid robot in space. In: Robotics and Automation (ICRA), 2011 IEEE International Conference on 2011, May. IEEE. pp. 2178-2183

[3] Chen Z, Lii NY, Wimböck T, Fan S, Liu H. Experimental evaluation of Cartesian and joint impedance control with adaptive friction compensation for the dexterous robot hand DLR-HIT II. International Journal of Humanoid Robotics. 2011;**8**(04): 649-671

[4] Sun W, Kong J, Wang X, Liu H. Innovative design method of the metamorphic hand. International Journal of Advanced Robotic Systems. 2018; **15**(1):1729881417754154

[5] Available from: http://es.bebionic.com/ [May 1, 2018]

[6] Azorin José M, et al. La Interacción de Personas con Discapacidad con el Computador: Experiencias y Posibilidades en Iberoamérica. Programa Iberoamericano de Ciencia y Tecnología para el Desarrollo (CYTED). 2013. ISBN-10: 84-15413-22-X

[7] Song Y, Mao J, Zhang Z, Huang H, Yuan W, Chen Y. A novel multi-objective shielding optimization method: DNN-PCA-NSGA-II. Annals of Nuclear Energy. 2021;**161**:108461

[8] Al-Fawa'reh M, Al-Fayoumi M, Nashwan S, Fraihat S. Cyber threat intelligence using PCA-DNN model to detect abnormal network behavior. Egyptian Informatics Journal. 2022;**23** (2):173-185

[9] Jiang N, Vujaklija I, Rehbaum H, Graimann B, Farina D. Is accurate mapping of EMG signals on kinematics needed for precise online myoelectric control? IEEE Transactions on Neural Systems and Rehabilitation Engineering. 2014;**22**(3):549-558

[10] Roche AD. Clinical perspectives in upper limb prostheses: An update. Current Surgery Reports. 2019;7:5. DOI: 10.1007/s40137-019-0227-z

[11] Hahne JM, Markovic M, Farina D. User adaptation in myoelectric man-machine interfaces. Scientific Reports. 2017;**7**(1):4437

[12] Hargrove LJ, Lock BA, Simon AM. Pattern recognition control outperforms conventional myoelectric control in upper limb patients with targeted muscle reinnervation. In: Proceedings of 2013 35th Annual International Conference of the IEEE Engineering in Medicine and Biology Society (EMBC). IEEE. 2013. pp. 1599–1602

[13] Wurth SM, Hargrove LJ. A real-time comparison between direct control, sequential pattern recognition control and simultaneous pattern recognition control using a Fitts' law style assessment procedure. Journal of Neuroengineering and Rehabilitation. 2014;**11**(1):91

[14] Kuiken TA, Miller LA, Turner K, Hargrove LJ. A comparison of pattern recognition control and direct control of a multiple degree-of-freedom transradial

prosthesis. IEEE Journal of Translational Engineering in Health and Medicine. 2016;**4**:1-8

[15] Chu JU, Moon YJ, Lee SK, Kim SK, Mun MS. A supervised feature-projection- based-real-time EMG pattern recognition for multifunction myoelectric hand control. IEEE Transaction on Mechatronics. 2007; **12**(3):282-290

[16] Chu J-U, Moon I, Mun M-S. A real-time EMG pattern recognition system based on linear-nonlinear feature projection for a multifunction myoelectric hand. IEEE Transactions on Biomedical Engineering. 2006;**53**:2232-2239

[17] Guler NF, Kocer S. Classification of EMG signals using PCA and FFT. Journal of Medical Systems. 2005;**29**(3): 29241-29250

[18] Smith RJ, Tenore F, Huberdeau D, Etienne-Cummings R, Thakor NV. Continuous decoding of finger position from surface EMG signals for the control of powered prostheses. In: Proceedings of 30th Annual International Conference of the IEEE EMBS. Vancouver, British Columbia. August 20–25, 2008

[19] Wang JZ, Wang RC, Li F, Jiang MW, Jin DW. EMG signal classification for myoelectric teleoperating a dexterous robot hand. In: Proceedings of 27th Annual International conference of the IEEE EMBS; Shanghai, China. January 17–18, 2006

[20] Kiatpanichagij K, Afzulpurkar N. Use of supervised discretization with PCA in wavelet packet transformation-based surface electromyogram classification. Biomedical Signal Processing and Control. 2009;**4**(2):127-138

[21] Hargrove L, Guangline L, Englehart K, Hudgins B. Principal Component's analysis preprocessing for improved classification accuracies in pattern-recognition-based myoelectric control. IEEE Transactions on Biomedical Engineering. 2019;**56**(5): 1407-1414

[22] Côté-Allard U, Fall CL, Drouin A, Campeau-Lecours A, Gosselin C, Glette K, et al. Deep learning for electromyographic hand gesture signal classification using transfer learning. IEEE Transactions on Neural Systems and Rehabilitation Engineering. 2019; **27**(4):760-771

[23] Amamcherla N, Turlapaty A, Gokaraju B. A machine learning system for classification of emg signals to assist exoskeleton performance. In 2018 IEEE Applied Imagery Pattern Recognition Workshop (AIPR). IEEE. October, 2018. pp. 1-4

[24] Boostani R, Moradi MH. Evaluation of the forearm EMG signal features for the control of a prosthetic hand. Physiological Measurement. 2003;**24**: 309-319

[25] Côté-Allard, Ulysse et al. Transfer learning for sEMG hand gestures recognition using convolutional neural networks. 2017 IEEE International Conference on Systems, Man, and Cybernetics (SMC) Banff Center, Banff, Canada. October 5-8, 2017

[26] Li C et al. PCA and deep learning based myoelectric grasping control of a prosthetic hand. Biomedical Engineering Online. 2018;**17**:107. DOI: 10.1186/ s12938-018-0539-8

[27] Wei W, Wong Y, Du Y, Hu Y, Kankanhalli M, Geng W. A multi-stream convolutional neural network for sEMG-based gesture recognition in muscle-computer interface. Pattern Recognition Letters. 2019;**119**:131-138

[28] Franti E et al. Methods of acquisition and signal processing for myoelectric control of artificial arms. Romanian Journal of Information Science and Technology. 2012;**15**(2):91-105

[29] Cognolato M, Atzori M, Marchesin C, Marangon S, Faccio D, Tiengo C, et al. Multifunction control and evaluation of a 3D printed hand prosthesis with the Myo armband by hand amputees. BioRxiv. 2018:445-460

[30] Díaz-Amador R, Mendoza-Reyes MA, Cárdenas-Barreras JL. Reducing the effects of muscle fatigue on upper limb myoelectric control using adaptive LDA. Ingeniería Electrónica, Automática y Comunicaciones. 2019;**40**(2):10-21

[31] Campbell E, Phinyomark A, Al-Timemy AH, Khushaba RN, Petri G, Scheme E. Differences in EMG feature space between able-bodied and amputee subjects for myoelectric control. In 2019 9th International IEEE/EMBS Conference on Neural Engineering (NER) IEEE. 2019. pp. 33-36

[32] Yang Z, Jiang D, Sun Y, Tao B, Tong X, Jiang G, et al. Dynamic gesture recognition using surface EMG signals based on multi-stream residual network. Frontiers in Bioengineering and Biotechnology, 2021;**9**

[33] Bao T, Zaidi SAR, Xie S, Yang P, Zhang ZQ. A CNN-LSTM hybrid model for wrist kinematics estimation using surface electromyography. IEEE Transactions on Instrumentation and Measurement. 2020;**70**:1-9

[34] Liu J, Chen W, Li M, Kang X. Continuous recognition of multifunctional finger and wrist movements in amputee subjects based on sEMG and accelerometry. The Open Biomedical Engineering Journal. 2016;**10**:101

[35] Atzori M, Gijsberts A, Castellini C, Caputo B, Hager AGM, Elsig S, et al. Electromyography data for non-invasive naturally-controlled robotic hand prostheses. Scientific Data. 2014;**1**(1):1-13

[36] Bird JJ, Kobylarz J, Faria DR, Ekárt A, Ribeiro EP. Cross-domain MLP and CNN transfer learning for biological signal processing: EEG and EMG. IEEE Access. 2020;**8**:54789-54801

[37] Akhlaghi N, Dhawan A, Khan AA, Mukherjee B, Diao G, Truong C, et al. Sparsity analysis of a sonomyographic muscle–Computer interface. IEEE Transactions on Biomedical Engineering. 2019;**67**(3):688-696

[38] Rabin N, Kahlon M, Malayev S, Ratnovsky A. Classification of human hand movements based on EMG signals using nonlinear dimensionality reduction and data fusion techniques. Expert Systems with Applications. 2020; **149**:113281

[39] Tsinganos P, Cornelis B, Cornelis J, Jansen B, Skodras A. Deep Learning in EMG-based Gesture Recognition. In: PhyCS. 2018. pp. 107-114

[40] Ramírez-Martínez D, Alfaro-Ponce M, Pogrebnyak O, Aldape-Pérez M, Argüelles-Cruz AJ. Hand movement classification using burg reflection coefficients. Sensors. 2019;**19**(3):475

[41] Dirgantara GP, Basari B. Optimized circuit and control for prosthetic arm based on myoelectric pattern recognition via power spectral density analysis. In AIP Conference Proceedings. AIP Publishing LLC. 2019;**2092**(1):020013

[42] Benatti S, Milosevic B, Farella E, Gruppioni E, Benini L. A prosthetic hand body area controller based on efficient

pattern recognition control strategies. Sensors. 2017;**17**(4):869

[43] Ortiz-Catalan M, Rouhani F, Branemark R, Hakansson B. Offline accuracy: A potentially misleading metric in myoelectric pattern recognition for prosthetic control. In: Proceedings of 2015 37th Annual International Conference of the IEEE Engineering in Medicine and Biology Society (EMBC). IEEE. 2015. pp. 1140–1143

[44] Wang S, Chen B. Split-stack 2D-CNN for hand gestures recognition based on surface EMG decoding. In: 2020 Chinese Automation Congress (CAC). IEEE. November 2020. pp. 7084-7088

[45] Côté-Allard U, Gagnon-Turcotte G, Laviolette F, Gosselin B. A low-cost, wireless, 3-D-printed custom armband for sEMG hand gesture recognition. Sensors. 2019;**19**(12):2811

[46] Hassan HF, Abou-Loukh SJ, Ibraheem IK. Teleoperated robotic arm movement using electromyography signal with wearable Myo armband. Journal of King Saud University-Engineering Sciences. 2020;**32**(6): 378-387

[47] Ozdemir MA, Kisa DH, Guren O, Onan A, Akan A.. EMG based hand gesture recognition using deep learning. In: 2020 Medical Technologies Congress (TIPTEKNO). IEEE. 2020, November. pp. 1-4

[48] Chen L, Fu J, Wu Y, Li H, Zheng B. Hand gesture recognition using compact CNN via surface electromyography signals. Sensors. 2020;**20**(3):672

[49] Tsinganos P, Cornelis B, Cornelis J, Jansen B, Skodras A. Data augmentation of surface electromyography for hand gesture recognition. Sensors. 2020; **20**(17):4892

Chapter 4

Perspective Chapter: Cooperation among Humans and Robots in Remote Robot Systems with Force Feedback

Pingguo Huang and Yutaka Ishibashi

Abstract

By using remote robot systems with force feedback, we can largely improve the efficiency and accuracy of work among the systems over a network. In such a system, a human can operate a remote robot by manipulating a haptic interface device while monitoring the movement of the robot arm and perceiving force applied to an object touched/moved by an arm of the robot having a force sensor. The remote robot systems with force feedback are expected to be used in many areas such as outer space, deep sea, nuclear power plants, and disaster areas, which humans cannot enter easily. In these situations, three types of cooperation among humans and robots (that is, between humans and robots, between humans, and between robots) are highly demanded. In this chapter, we introduce our remote robot systems with force feedback and describe the three types of cooperation among humans and robots in the systems. We also explain QoS (Quality of Service) control and stabilization control as our challenges and solutions for effective cooperation. Furthermore, we discuss future directions of the cooperation.

Keywords: remote robot system, force feedback, haptic interface device, cooperation, QoS control, stabilization control

1. Introduction

In recent years, many researchers pay attention to remote robot systems with force feedback [1–6]. In such a system, a human can operate a remote robot by manipulating a haptic interface device while monitoring movement of the robot arm with a video camera and perceiving force from an object touched/moved by an arm of the robot having a force sensor. Since the user can touch and feel the shape, weight, and softness of the object, the efficiency and accuracy of remote operation can be largely improved [7]. Therefore, the remote robot systems with force feedback are able to be used in various areas such as outer space, deep sea, nuclear power plants, and disaster areas, in which humans cannot enter easily. In these applications, it is important to make cooperation among humans and robots with force feedback efficiency. We focus on the cooperation among humans and robots with force feedback in this chapter.

IntechOpen

When the control information (i.e., force and/or position) is transmitted over a network that does not guarantee QoS (Quality of Service) [8], owing to network problems such as network delay, delay jitter, and packet loss, the QoE (Quality of Experience) [9] may be seriously deteriorated. Also, unstable phenomena such as vibrations may occur [10–12], and cooperation may not be well. To solve the problems, we need to carry out QoS control and stabilization control. QoS control alleviates the influences of network delay, delay jitter, and packet loss, and stabilization control suppresses instability phenomena such as vibrations of the robot and haptic interface device. We mainly focus on the problems and their solutions for effective cooperation among humans and robots with force feedback.

In this chapter, first, we explain the cooperation among humans and robots with force feedback in Section 2. Next, we introduce the remote robot systems with force feedback in Section 3. Then, we outline the problems to be solved in the cooperation among humans and robots with force feedback in Section 4 and explain our solutions for effective cooperation in Section 5. Finally, we discuss the future directions of the cooperation among humans and robots in Section 6 and we conclude the chapter in Section 7.

2. Cooperation among humans and robots with force feedback

In a remote robot system with force feedback, a human can operate a haptic interface device to control/support a remote robot while watching the video received from the remote terminal. From the haptic interface device, position information which is used to control/support remote robot is transmitted to the robot, and the position information of robot arm and force information sensing by force sensor attached to the robot is transmitted to the haptic interface device for outputting reaction force (see **Figure 1**, in which a toggle clamp hand is attached to the force sensor. We can attach various types of hands).

The remote robot systems with force feedback can be used for cooperation among humans and robots. **Figure 2** shows examples of cooperation between the two remote robots. In **Figure 2(a)**, the two remote robots collaboratively carry and move an object (wooden stick) from one position (initial position) to another position (destination). In **Figure 2(b)**, one robot (robot arm 1) moves an object (wooden stick) toward another robot (robot arm 2), and hands over the object to robot arm 2. In **Figure 2(c)**, the two remote robots carry an object (cardboard box) by putting the object from both sides and moving the object from one position to another position. Cooperation can be widely used in many areas, and several types of cooperation exist. As shown in **Figure 3**, the cooperation among humans and robots can be grouped into cooperation between human-human, human–robot, and robot–robot. In this section,

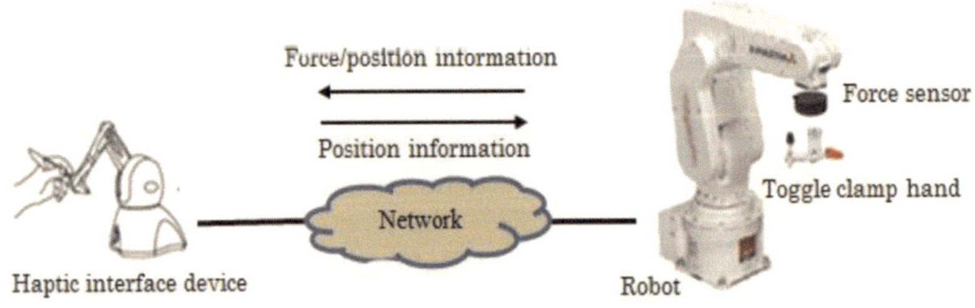

Figure 1.
Remote robot system with force feedback.

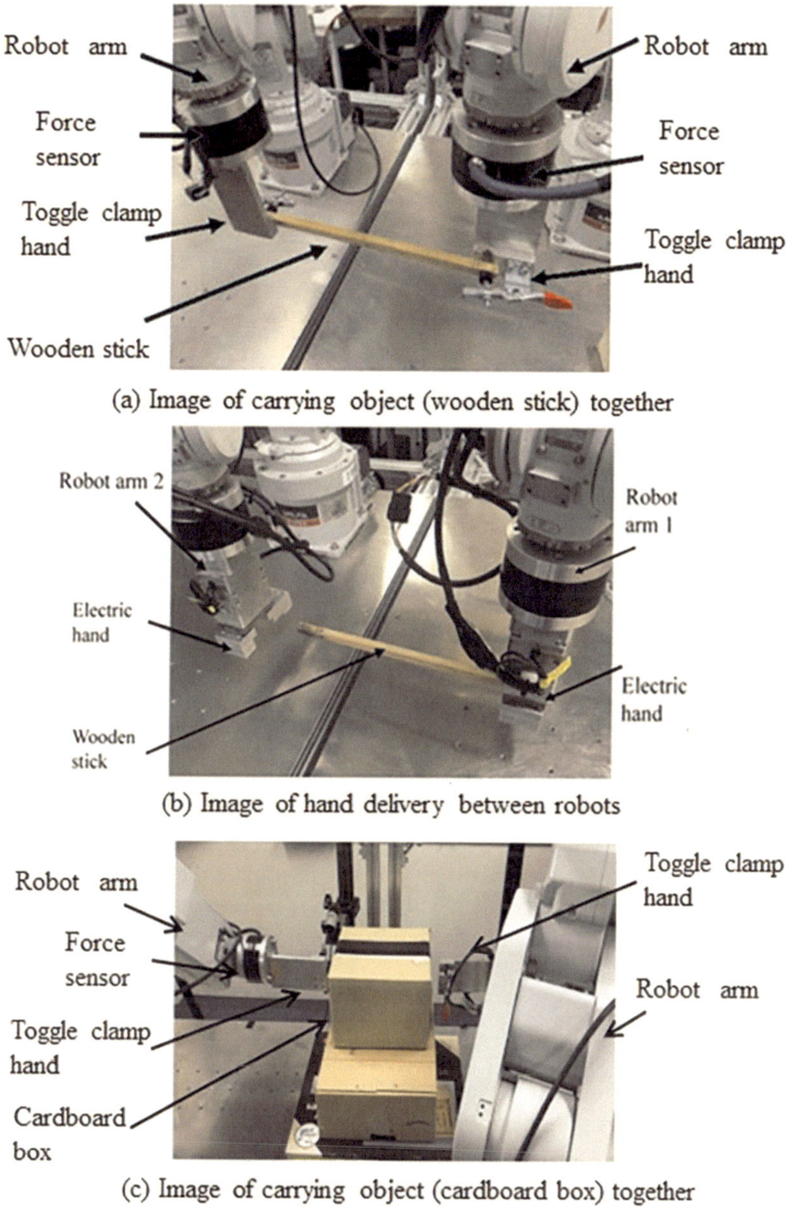

Figure 2.
Cooperation images between two robots.

we introduce expected applications using the remote robot systems with force feedback and explain the types of cooperation.

2.1 Expected applications

The cooperation among humans and robots in the remote robot systems with force feedback is expected to be used in various areas such as dangerous areas in which humans cannot enter easily (for example, outer space, deep sea, and

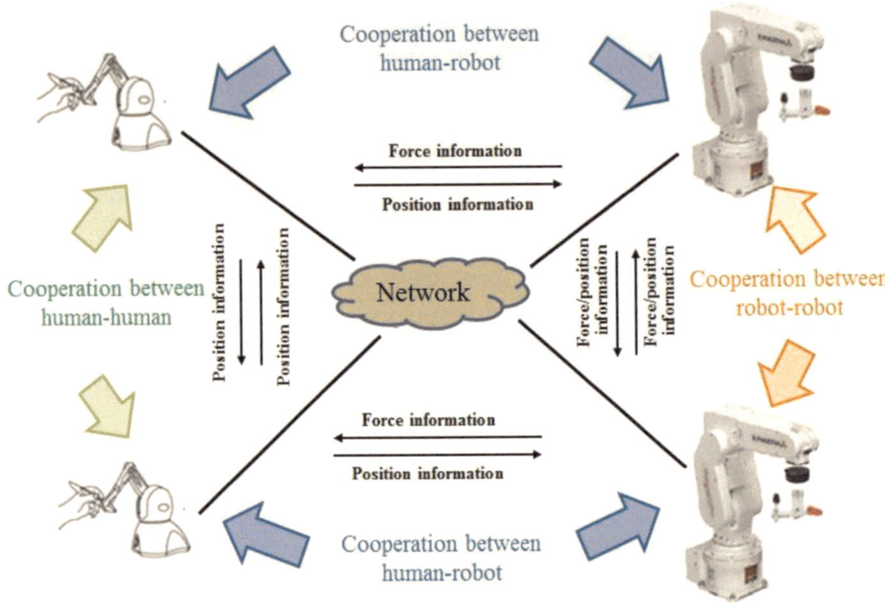

Figure 3.
Cooperation among humans and robots.

radioactively contaminated areas), disaster areas (disaster rescue and relief), and remote medicine.

2.1.1 Applications in dangerous areas

It is dangerous for humans to work in dangerous areas such as outer space, deep sea, and radioactively contaminated areas. We can use robots to work in such areas instead of humans entering the areas while the humans control the robots in safe areas.

2.1.2 Disaster areas (rescue and relief)

After the outbreak of disasters such as earthquakes and concentrated downpours, rescue and relief are the most important. However, it may be difficult for humans to enter the hard-hit areas. We can control remote robots to confirm disaster situations for making effective rescue plans. Also, humans in safe areas can control remote robots to enter the areas for rescue and relief.

2.1.3 Remote medicine

There exist large health disparities in different regional/national areas. Using remote robot systems with force feedback may be an effective method to solve the problem. The systems can be used for remote surgery and remote rehabilitation. Also, they can be employed for remote surgery training for medical interns.

2.1.4 Other areas

The remote robot systems with force feedback are also expected to be used for various applications such as industrial factories and home delivery in other areas. For industrial factories, humans can control remote robots to enter extreme environments (for example, high temperature and/or high-pressure environments) or limited working areas which are difficult for humans for production work or inspection and maintenance. For homedelivery, humans can remote control drones or movable robots to deliver packages.

In these applications, it is difficult to conduct work with only robots if the situations and environments are unknown in advance and may be always changing. Therefore, human's support is needed [6]. This means that we need robots to help humans, and robots also need human's support. That is, in the applications, cooperation between a human (local user) and a robot (remote robot) in each system, that between humans, that between robots (remote robots), and that among humans and robots (remote robots) are needed. We explain the three types of cooperation in the next subsection.

2.2 Types of cooperation

As described in the previous subsection, robots can help humans, and robots also need humans' support. We explain the cooperation among humans and robots in this subsection.

1. Cooperation between human and robot

   In this type of cooperation, a human operates a haptic interface device to control/support a remote robot. Instead of the human, a robot does the work at a remote side and sends back the information about reaction force sensed by force sensor, and the information about a remote environment by other sensors such as video cameras and microphones. The information can help the human to know the remote situation. Also, it is difficult for the robot to do work in a complex environment independently. Therefore, human's support is needed for the robot. The human can control or adjust the movement of the remote robot to help the robot conduct the work.

2. Cooperation between robots

   In dangerous areas or disaster areas, wireless communication can be an effective method for communication between humans and robots. However, wireless communication is largely affected by weather, obstacles, and distances; this means that large network delays or network troubles may occur. These problems may make humans difficult to control or support remote robots to deal with abrupt changes in robots' positions. In this case, since humans require time to support the remote robots, cooperation between robots using force sensors is needed for quick response.

3. Cooperation between humans

   In this cooperation, multiple humans operate haptic interface devices to control remote robots to do work. In this case, to do the work smoothly and effectively, cooperation between humans is important. In cooperation, humans need to start to operate remote robots at the same time at the beginning of the cooperation work, keeping almost the same movement speed in the cooperation. Also, when multiple humans operate the remote robots to move an object, efficient cooperation may be needed for position's fine adjustment to the goal position.

Therefore, in cooperation, humans need to transmit their wills (e.g., the movement direction and movement speed) to each other for efficient cooperation.

4. Cooperation among humans and robots

For work that cannot be conducted by only one remote robot, multiple robots may be needed. The robots should be operated by multiple humans. In this case, we need the above three types of cooperation.

3. Remote robot systems with force feedback

In this section, we introduce the remote robot systems with force feedback constructed in our study.

3.1 System configuration

The configuration of the remote robot systems with force feedback is shown in **Figure 4**. Each system consists of a master terminal and a slave terminal. Each terminal has two PCs; that is, PC for haptic interface device and PC for video at the master terminal, and PC for industrial robot and PC for video (web camera) at the slave terminal. At the master terminal, a haptic interface device (3D Systems Touch [13]) having 3 DoF (Degree of Freedom) is connected to PC for haptic interface device. At the slave terminal, an industrial robot is connected to PC for industrial robot, and a web camera (produced by Microsoft Corp., and the video resolution is 1920 × 1080 pixels) is connected to PC for video. The industrial robot has a 6 DoF robot arm (RV-2F-D [14] by Mitsubishi Electric Crop.), a robot controller (CR750-Q [14]), and a force sensor (1F-FS001-W200 [15]) attached to the tip of the robot arm. The force sensor is connected to the robot controller through a force interface unit

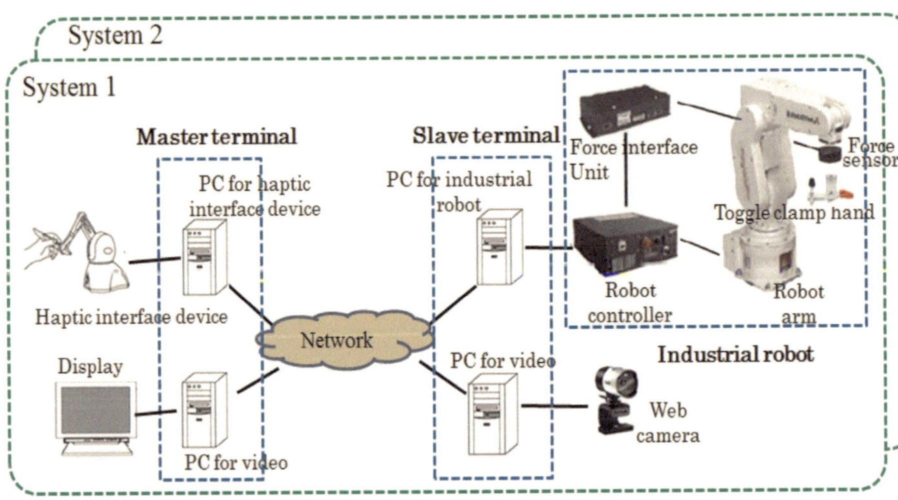

Figure 4.
Configuration of remote robot systems with force feedback.

(2F-TZ561 [15]). Also, a toggle clam hand/a electric hand is further linked to the force sensor.

In each system, the master terminal inputs the position information of the haptic interface device and calculates and outputs the reaction force for the haptic interface device every millisecond. The master terminal also transmits the position information to the slave terminal by UDP (User Datagram Protocol). At the slave terminal, based the position information received from the master terminal, a position vector is calculated and sent to the industrial robot every 3.5 milliseconds by the real-time control function [16]. The force information sensed by the force sensor is transmitted to the master terminal, and the master terminal calculates the reaction force outputted by the haptic interface device based on the received force information every millisecond as described previously.

At the master terminal, the reaction force $\boldsymbol{F}_t^{(\mathrm{m})}$ outputted through to the haptic interface device at time t ($t \geq 1$) is calculated as follows:

$$\boldsymbol{F}_t^{(\mathrm{m})} = \boldsymbol{K}_{\mathrm{scale}}^{(\mathrm{F})} \boldsymbol{F}_{t-1}^{(\mathrm{s})} \tag{1}$$

where $\boldsymbol{K}_{\mathbf{scale}}^{(\mathbf{F})}$ is a force scale which is set to 1 in this chapter, and $\boldsymbol{F}_{t-1}^{(\mathrm{s})}$ denotes the force received from the slave terminal (note that we use only the three axes (the x-, y-, and z- aexs) of force here). Furthermore, since the maximum force outputted through the haptic interface device is 3.3 N [13], the reaction force is set to 3.3 N when the calculated force is larger than 3.3 N.

At the slave terminal, the position vector $\boldsymbol{S}_t$ of the industrial robot outputted at the time t ($t \geq 2$) is calculated as follows:

$$\boldsymbol{S}_t = \boldsymbol{K}_{\mathrm{scale}}^{(\mathrm{P})} \boldsymbol{M}_{t-1} + \boldsymbol{V}_{t-1} \tag{2}$$

where $\boldsymbol{K}_{\mathbf{scale}}^{(\mathbf{P})}$ is a coefficient for mapping of workspace, $\boldsymbol{M}_t$ is the position vector of the haptic interface device received from the master terminal at time t, $\boldsymbol{V}_t (= \boldsymbol{S}_t - \boldsymbol{S}_{t-1})$ is the velocity vector of the industrial robot, and $|\boldsymbol{V}_t|$ $\boldsymbol{V}_{\mathbf{max}}$, where $\boldsymbol{V}_{\mathbf{max}}$ is the maximum movement velocity ($\boldsymbol{V}_{\mathbf{max}}$ = 5 mm/s [16] in this chapter) to operate the robot arm safely.

3.2 Cooperation methods

The remote robot systems with force feedback are used for serval types of cooperation, and we here introduce the cooperation methods.

3.2.1 Human-robot

As described in Subsection 2.2, the cooperation between human and robot, a human operates a haptic interface device to control/support a remote robot. In this type of cooperation, a single remote robot system is used, and the human can feel the reaction force sensed by the force sensor attached to the remote robot arm. By feeling the reaction force and watching video received from remote side, the human can support/control the remote robot by sending position information which is used to move the remote robot.

We conduct the cooperation between human-robot in our previous work [12, 17]. In Ref. [12], the authors handle work of pushing a ball and identifying what kind of ball is pushed according to the softness without watching video. In Ref. [17], a human can manipulate a robot arm with a pen to write characters by using a haptic interface device, and the authors make a comparison among different types of stabilization control which we will explain in Section 5.

3.2.2 Robot-robot

As described previously, the remote robot systems can be used for the area such as disaster areas and deep see, in which communication failure, and unforeseen events and abrupt change of position may occur frequently. In this case, robots cannot receive the support/control information from humans. The cooperation between robot and robot can be used to solve the problem by working independently without humans' support. Also, the cooperation can be used under the ordinary environment. In order to conduct the cooperative work independently, the robots can cooperate with each other according to the force information sensed by force sensors and/or by transmitting position information transmitted between the robots.

In Ref. [18], the authors deal with cooperation work in which an object (a wooden stick) is carried together by using the two remote robot systems with force feedback. In the cooperation, according to the force sensed by force sensor, one robot adjusts the position in the direction of reducing the force applied to the object to cooperate with another robot. We also suppose that mobile robots suddenly move up and down largely and employed the enhanced control of the robot position control using force information to against sudden and large position change in the vertical direction [19]. We further investigate the effects of the enhanced control by employing one axis among the six axes for the robot movement, and we regard the industrial robots as mobile robots (i.e., pseudo-mobile robots). In Ref. [18], the two remote robot systems are used to do collaborative work of carrying an object (a wooden stick) together. In the cooperation, each slave terminal (robot side) also sends the position information of robot arm to that of the other system (robot side). In Ref. [20], two humans operate the two remote robot systems to do cooperative work of carrying an object. In the cooperation, the position information is transmitted between robots to make the cooperation smooth.

3.2.3 Human-human

When multiple humans control remote robot systems with force feedback to do work, the cooperation between humans is important. In this type of cooperation, communication between humans is necessary. Humans can transmit their wills (e.g., the movement direction and movement speed) to each other by using sound, video, and/or force information (i.e., position information which can be used to calculate the reaction force outputted through a haptic interface device) and make the cooperation smooth.

In Ref. [21], one of two humans operate a remote robot to carry an object together with the other human by using a haptic interface device. In the cooperation, the position information of the haptic interface device is transmitted to calculate the reaction force outputted through haptic interface device. In Ref. [22], one human operates a haptic interface device to control a remote robot to carry and move an object (a wooden stick), and hand over the object to another remote robot which is controlled by another human.

3.2.4 Humans-robots

For work which cannot conduct successfully by only one remote robot system, multiple remote systems with force feedback and multiple humans may be needed. When multiple remote robot systems with force feedback are employed to do collaborative work, cooperation among humans and robots is needed. That is, communications between human and robot, between robot and robot, and between human and human are required.

4. Problems to be solved

In order to achieve effective cooperation, many problems should be solved. In this section, we explain the problems by grouping them into QoS problems and stabilization problems.

4.1 QoS problems

As described in Section 1, when the control information (i.e., force and/or position) is transmitted over a network which does not guarantee QoS like the Internet, owing to the network problems such as the network delay, delay jitter, and packet loss, the following problems (e.g., QoE deterioration and poor cooperation) may occur.

1. Fluctuation of reaction force

   As the network delay increases, the reaction force outputted through the haptic interface device may become larger. That is, it is more difficult for a human to operate the device.

2. Out of temporal synchronization (known as media synchronization [23])

   Owing to the network delay and delay jitter, media synchronization (i.e., the temporal relationship of media units (MUs), which are information units for media synchronization [24]) problems may occur. The problems of temporal synchronization can be grouped into problems of intra-stream synchronization, inter-stream synchronization, and inter-destination (or group) synchronization [23].

3. Out of spatial synchronization

   In the cooperation, robots need to move/operate an object at the same altitude, same angle, and same speed. However, humans and robots are at different locations, out of spatial synchronization (i.e., robots move/operate the object in different altitude, different angle, and different speed) may occur, and cooperation may not be well. This is because if spatial asynchrony occurs, the force applied to the operating object becomes large, and the object may be damaged by the large force. That is, we need to keep temporal and spatial synchronization high.

   To solve the problems, we need to carry out QoS control, and we will introduce the QoS control in Subsection 5.1.

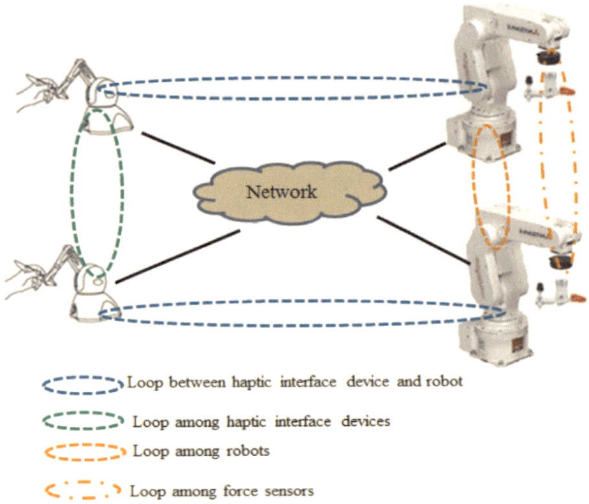

Figure 5.
Control loops in remote robot systems with force feedback.

4.2 Stabilization problems

In the remote robot system with force feedback, as the network delay increases, the reaction force becomes larger, and instability phenomena such as vibrations of the robot and haptic interface device may occur more frequently [12]. Furthermore, when multiple remote robot systems are used for cooperation, there are multiple loops caused by communication and force sensors in the systems (see **Figure 5**), and there exist close relationships among loops. Therefore, instability phenomena may become more often and more complex.

5. Solutions for effective cooperation

In this section, we introduce solutions which can be used to solve the problems described in Section 4 for effective cooperation.

5.1 QoS control

In order to solve the QoS problems, we need to carry out QoS control. Various types of QoS control have been proposed [5]. We explain serval types of QoS control which can be used to solve the QoS problems for the cooperation among remote robot systems.

1. Temporal synchronization control

   Temporal synchronization control can be used to solve the problems caused by the network delay and delay jitter [23] and following types of control are included. In the control, we suppose that the global synchronization clocks [25], which have the same value and advancement, are used at all the terminals for simplicity. We can adjust the clocks by Network Time Protocol (NTP) [26] or Global Positioning System (GPS) easily.

a. Intra-stream synchronization control

   The control is used to keep the timing relation between MUs such as video pictures and voice packets in a single media stream. Serval types of intra-stream synchronization control such as Skipping [24], Virtual-Time Rendering (VTR) [24] have been proposed.

   In the remote robot systems with force feedback, the control can be used to keep the timing relation of each media stream such as haptic/force media and video for each system.

b. Inter-stream synchronization control

   The inter-stream synchronization control is used to preserve the temporal relationship among multiple media streams. VTR can also be used for inter-stream synchronization control. Under the control, one media stream is selected as the master stream, and the others are handled as slave streams. For the master stream, we carry out only the intra-stream synchronization control, and we exert the inter-stream synchronization control which adjusts the output timing of each slave stream to that of the master stream after carrying out the intra-stream synchronization control.

   In the remote robot systems, haptic/force streams and video may be transmitted between the master and slave terminals in the cooperation between human-robot/humans-robots; the haptic stream and voice may be transmitted among the master terminals (i.e., humans' sides) in the cooperation of humans-humans. In these cases, multiple streams are transmitted among the terminals, and we need to carry out the inter-stream synchronization control. Normally, the most important stream such as the haptic/force stream can be selected as the master stream under the control.

c. Inter-destination synchronization control

   The inter-destination synchronization control can be employed to output each MU simultaneously at different terminals (destinations). When multiple remote robot systems with force feedback are used to do collaborative work, if each MU cannot be outputted simultaneously at different terminals, the cooperation may not be well because humans may watch different displayed video images of cooperation work and feel different reaction force. Therefore, we need to carry out the inter-destination synchronization control to improve the efficiency of cooperation work. Also, the control can be used for spatial synchronization.

2. Adaptive reaction force control (adaptive elasticity control)

   In each remote robot system with force feedback, the reaction force of the haptic interface device is normally calculated from the force sensed by force sensor (see Eq. (1)). When the master terminal receives position information from the slave terminal, the spring-damper model [27] can be used to calculate the reaction force. In the spring-damper model, the reaction force includes the elasticity and viscosity. The elasticity is force exerted by deformation of a spring or rubber. For example, when a spring is pushed or pulled, the elasticity is proportional to the depth of the spring, and it is calculated by multiplying the depth by the elastic spring coefficient. The viscosity is force or resistance exerted by fluids. For example, when we move an object in the fluids (e.g., water and oil). The viscosity is proportional to the relative velocity (i.e., the velocity of the object relative to the fluids), and it can be calculated by multiplying the relative velocity by the viscosity (damper) coefficient. The reaction force is calculated as follows:

$$\boldsymbol{F}_t^{(\mathrm{m})} = -K_{\mathrm{s}}\left(\boldsymbol{P}_{t-1}^{(\mathrm{m})} - \boldsymbol{P}_{t-1}^{(\mathrm{s})}\right) - K_{\mathrm{d}}\left(\boldsymbol{v}_{t-1}^{(\mathrm{m})} - \boldsymbol{v}_{t-1}^{(\mathrm{s})}\right) \tag{3}$$

where K_{s} is the spring coefficient, K_{d} is the damper coefficient, $\boldsymbol{P}_{t-1}^{(\mathrm{m})}$ is the position vector of the haptic interface device at the master terminal, $\boldsymbol{P}_{t-1}^{(\mathrm{s})}$ is that at the slave terminal, $\boldsymbol{v}_{t-1}^{(\mathrm{m})}$ is the velocity of the haptic interface device at the master terminal, and $\boldsymbol{v}_{t-1}^{(\mathrm{s})}$ is that at the slave terminal. By using this model, a human can still control remote robot because the human can perceive the reaction force even if other parts of the robot with force sensors hit an object and do not move further. Note that the sensors do not sensed force in this case.

The adaptive reaction force control [7] includes the adaptive viscosity control [28], adaptive elastic control [29], and adaptive viscoelasticity control [30]. The adaptive elastic control dynamically changes the elastic coefficient according to network delay. The adaptive viscosity control dynamic changes the viscosity coefficient according to the network delay and the velocity of the haptic interface device. The adaptive viscoelasticity control exerts the two types of control.

3. Robot position control using force information

   The robot position control using force information [31] is proposed to reduce the force applied to an object carrying by two remote robot systems in cooperative work of carrying the object. The control finely adjusts the position of the robot arm in the direction which leads to small reaction force.

4. Adaptive Δ-causality control

   The adaptive Δ-causality control dynamically changes the value of Δ which is used for outputting each MU simultaneously (i.e., the MU is output at the time limit of generation time plus Δ seconds) according to the network load (i.e., the network delay and delay jitter) [7]. If the MU is received after the time limit, it is discarded as an obsolete MU. In the remote robot system with force feedback, the control can be used to output of the position information simultaneously among multiple terminals.

   In [20], the authors propose the global Δ-causality control and compare the proposed control with the local adaptive Δ-causality control by experiment. The local adaptive Δ-causality control is partially applied to the remote robot systems (i.e., the adaptive Δ-causality control is carried out between the two robots, or the control is carried out in each remote robot system). To output each MU simultaneously at each system, we need to carry out the global adaptive Δ-causality control in which the control is globally performed (i.e., the adaptive Δ-causality control is carried out between the two robots and in each remote robot system). Experimental results show that the global control is more effective than the local control for cooperation.

The control can also be used for spatial synchronization.

5. Force adjustment control

The force adjustment control is proposed for cooperative work (see **Figure 2(c)**) of carrying an object by putting the object from both sides between two remote robot systems [32], and the control is used to suppress large force applied to the object and to avoid dropping it because of too small force. When the force applied to the object carried by the two robot arms is larger than a threshold value, each robot arm moves in the direction to reduce the force by a certain distance. When the force is smaller than another threshold value, the robot arms adjust in the direction to increase the force.

Based on the characteristics of the above QoS control, we can summarize the relationship between QoS control and the three types of cooperation in **Table 1**.

For the temporal synchronization control, the intra-stream synchronization control is needed for each terminal in the cooperation of human-robot, robot-robot, and human-human; the inter-stream synchronization control is required for the cooperation of human-robot in which multiple streams are transmitted between the master and slave terminals; the inter-destination synchronization control is necessary for the cooperation of robot-robot and human-human, in which the same types of operation are needed at the terminals.

The adaptive reaction force control can be used to reduce the large reaction force caused by network delay in the cooperation of human-robot (i.e., position difference which is used to calculate the reaction force becomes large as network delay increases) and human-human.

The robot position control using force information can be used to reduce the force applied to an object carrying by two robots in the cooperation of human-robot and robot-robot.

The adaptive Δ-causality control can be employed to alleviate the influence of difference in network delay between the haptic interface device and robot in the

	Human-robot	Robot-robot	Human-human
Temporal synchronization control	○	○	○
Adaptive reaction force control	○	—	○
Robot position control using force information	○	○	—
Adaptive Δ-causality control	○	○	—
Force adjustment control	○	○	—

Table 1.
Close relationship between QoS control and cooperation.

cooperation of human-robot and can be used to reduce the influence of difference in network delay between the robots in the cooperation of robot-robot.

The force adjustment control can be used to adjust the force applied to the operating object in suitable range of values, and the control can be used for the cooperation of human-robot and robot-robot.

For the cooperation of humans-robots, since the cooperation including the three types of cooperation, all the above types of control can be utilized.

5.2 Stabilization control

In order to achieve stable and high quality of the cooperation using the remote robot systems with force feedback, we need to carry out stabilization control as well as QoS control. This subsection introduces serval types of stabilization control which are employed in our remote robot systems.

1. Stabilization control with filters

   The stabilization control with filters uses the wave filter in combination with the phase control filter [12]. The stabilization control with filters makes the remote robot system with force feedback stable against large network delays.

2. Stabilization control by viscosity

   In the stabilization control by viscosity [33], the instability phenomenon is suppressed by using viscosity. The control uses the coefficient to restrict the movement distance of the industrial robot to some extent compared with the movement distance of the haptic interface device.

3. Reaction force control upon hitting

   In order to solve the vibration problem when the remote robot hits/touches a hard object, the reaction force control upon hitting [34] is proposed. The reaction force control upon hitting gradually increases the reaction force to prevent a robot arm from jumping when the arm hits a hard object (i.e., when the reaction force calculated based on the force information received from the force sensor attached to the remote robot arm is larger than a threshold value).

4. Stabilization control by viscosity

   Since the viscosity can be used to suppress vibration, the adaptive viscosity control can also be used as stabilization control. In the control, viscosity is generated by decreasing the movement distance of the industrial robot by a certain amount proportional to the movement velocity [33].

6. Future directions

In the previous section, we have introduced serval types of QoS control and stabilization control to solve the problems in the cooperation among humans and robots. However, there still exist many challenges. In this section, we discuss the future directions of QoS control and stabilization control for further stable and efficient cooperation.

1. Integrated QoS control

   In the cooperation using multiple remote robot systems with force feedback, we need to carry out multiple types of QoS control for the systems. If we carry out the QoS control independently, QoE may be deteriorated owing to excessive or insufficient effects of the control [7]. Therefore, we need to carry out different types of QoS control in an integrated manner like the adaptive QoS control [35]; for example, we need to carry out error control [5] together with traffic control [5] to improve the efficiency. Note that the error control increases the network traffic volume. In such a way, if we perform the QoS control independently, excessive or insufficient effects of the control may occur.

2. Global QoS control

   In the cooperation among multiple remote robot systems, it is important to carry out QoS control for each system and/or each terminal. However, it is also necessary to carry out QoS control for the multiple systems globally (i.e., we need to take account the whole systems when we carry out the QoS control). This is because multiple systems are used to conduct one work together. In Ref. [20], as described in Subsection 5.1 (4), we also found that the global adaptive Δ-causality control is more effective than the local adaptive Δ-causality control in the cooperation between the two remote robot systems.

3. Synthesis of QoS control and stabilization control

   Since we need to carry out QoS control and stabilization control for remote cooperation using multiple remote robot systems, it is important to integrate the QoS control with stabilization control to achieve stable and high quality of systems. We can integrate QoS control into stabilization control for the system. For example, we can carry out QoS control in the loop of stabilization control [6].

4. Application of AI

   Many factors such as contents of work, movement speed, network delay, room temperature, and wind may affect the effects of QoS control and stabilization control [36]. In order to achieve efficient cooperation (stable and high QoS), big data [37], cloud computing [38], and AI (Artificial Intelligence) [39] technologies such as neural network, fuzzy theory, and genetic algorithm can be useful methods for taking account of the factors.

7. Conclusions

In this chapter, we focused on the cooperation among humans and robots in remote robot systems with force feedback. We explained the expected applications and the four types of cooperation by using the remote robot systems. We also introduced our remote robot systems with force feedback and how to cooperate among the systems. We further explained QoS (Quality of Service) control and stabilization control as our challenges and solutions for effective cooperation. Furthermore, we discussed future directions of the cooperation.

Acknowledgements

The authors thank Prof. Takanori Miyoshi of Nagaoka University of Technology, Prof. Takashi Okuda of Aichi Prefectural University, Prof. Hitoshi Ohnishi of the Open University of Japan, and Prof. Hitoshi Watanabe of Tokyo University of Science for their valuable discussions.

Conflict of interest

The authors declare no conflicts of interest associated with this chapter.

Author details

Pingguo Huang[1]* and Yutaka Ishibashi[2]

1 Gifu Shotoku Gakuen University, Gifu, Japan

2 Nagoya Institute of Technology, Nagoya, Japan

*Address all correspondence to: huangpg@gifu.shotoku.ac.jp

References

[1] Ohnishi K. Real world haptics: Its principle and future prospects. The Journal of the Institute of Electrical Engineers of Japan. 2013;**133**:268-269 (In Japanese)

[2] Music S, Salvietti G, Dohmann P, Chinello F, Prattichizzo D, Hirche S. Human–robot team interaction through wearable haptics for cooperative manipulation. IEEE Transactions on Haptics. 2019;**12**(3):350-362

[3] Singh J, Srinivasan A, Neumann G, Kucukyilmaz A. Haptic-guided teleoperation of a 7-DoF collaborative robot arm with an identical twin master. IEEE Transactions on Haptics. 2020;**13**(1):246-252

[4] Haruna M, Kawaguchi N, Ogino M, Akino T. Comparison of three feedback modalities for haptics sensation in remote machine manipulation. IEEE Robotics and Automation Letters. 2021;**6** (3):5040-5047

[5] Ishibashi Y, Huang P. Improvement of QoS in haptic communication and its future. The IEICE Transactions on Communications. 2016;**J99-B**:911-925 (In Japanese)

[6] Huang P, Ishibashi Y. QoS control in remote robot operation with force feedback. In: Robotics Software Design and Engineering. London, UK: IntechOpen; 2020. pp. 1-13

[7] Huang P, Ishibashi Y. QoS control and QoE assessment in multi-sensory communications with haptics. The IEICE Transactions on Communications. 2013; **E96-B**:392-403. DOI: 10.1587/transcom. E96.B.392

[8] ITU-T Rec. I. 350 General Aspects of Quality of Service and Network Performance in Digital Networks. 1993

[9] ITU-T Rec. G. 100/P. 10 Amendment 1, New appendix I - Definition of quality of experience (QoE). 2007

[10] Miyoshi T, Maeda Y, Morita Y, Ishibashi Y, Terashima K. Development of haptic network game based on multi-lateral tele-control theory and influence of network delay on QoE. Transactions of the Virtual Reality Society of Japan, Special Issues on Haptic Contents. 2014; **19**:559-569 (In Japanese)

[11] Qian Q, Toyoda Y, Ishibashi Y, Huang P, Tateiwa Y. Effect of stabilization control on cooperative work between remote robot systems with force feedback. International Journal of Mechanical Engineering and Robotics Research (IJMERR). 2020;**9**(3):415-420

[12] Huang P, Miyoshi T, Ishibashi Y. Enhancement of stabilization control in remote robot system with force feedback. International Journal of Communications, Network and System Sciences (IJCNS). 2019;**12**(7):99-111

[13] https://www.3dsystems.com/haptics-devices/touch

[14] http://www.mitsubishielectric.co.jp/fa/products/rbt/robot/lineup/manual/f/bfp-a8899k.pdf (In Japanese)

[15] http://dl.mitsubishielectric.co.jp/dl/fa/members/document/manual/robot/bfp-a8940/bfp-a8940Z.pdf (In Japanese)

[16] http://dl.mitsubishielectric.co.jp/dl/fa/members/document/manual/robot/bfp-a8080/bfp-a8080e.pdf (In Japanese)

[17] Ye R, Ishibashi Y, Huang P, Tateiwa Y. Comparison of stabilization control for writing characters in remote robot system with force feedback. In:

Proc. the 3rd International Conference on Computer Communication and the Internet (ICCCI), Nagoya, Japan. 2021

[18] Ito S, Ishibashi Y, Huang P, Tateiwa Y. Effect of robot position control using force information: Human versus robot with force sensor. In: Proc. the 9th International Conference on Information and Education Technology (ICIET), Okayama, Japan. 2021. pp. 257-261

[19] Ishibashi Y, Huang P, Psannis KE. Enhanced robot position control using force information for mobile robots: Influences of obstacles on cooperative work. In: Proc. 2021 IEEE 9th International Conference on Information, Communication and Networks (ICICN), the 4th World Symposium on Communication Engineering (WSCE) Held as Workshop of ICICN, Xi'an, China. 2021. pp. 555-559

[20] Hameedha N, Ishibashi Y. Effects of local and global adaptive Δ-causality control on cooperative work between remote robot systems with force feedback. ITE Transactions on Media Technology and Applications (MTA). 2022;**10**(1):1-7

[21] Ye R, Ishibashi Y, Huang P, Tateiwa Y. Comparison of collaboration methods between users in remote robot systems with force feedback. In: Proc. the 7th International Conference on Computer and Communications (ICCC), Chengdu, China. 2021. pp. 1052-1056

[22] Toyoda Y, Ishibashi Y, Huang P, Tateiwa Y, Watanabe H. Efficiency of cooperation between remote robot systems with force feedback: Comparison with cooperation between user and remote robot system. International Journal of Mechanical Engineering and Robotics Research (IJMERR). 2020;**9**(6):900-905

[23] Ishibashi Y, Tsuji A, Tasaka S. A group synchronization mechanism for stored media in multicast communications. In: Proc. 16th IEEE International Conference on Computer and Communications (INFOCOM), Kobe, Japan. 1997. pp. 693-701

[24] Ishibashi Y, Tasaka S, Hasegawa T. The virtual-time rendering algorithm for haptic media synchronization in networked virtual environments. In: Proc. the 16th International Workshop on Communications Quality and Reliability (CQR), Okinawa, Japan. 2002. pp. 213-217

[25] Huang P, Ishibashi Y. Simultaneous output-timing control in networked games and virtual environments. In: Handbook on Multimedia Synchronization. Switzerland: Springer International Publishing; 2018. pp. 149-166

[26] Mills DL. Internet time synchronization: The network time protocol. IEEE Transactions on Communications. 1991;**39**(10):1482-1493

[27] 3D systems, OpenHaptics toolkit programmer's guide. version 3.2, 2013

[28] Komatsu Y, Ohnishi H, Ishibashi Y. Adaptive control of viscosity in remote control system with force feedback. In: Proc. IEEE International Conference on Consumer Electronics - Taiwan (ICCE-TW), Chinese Tapei. 2017. pp. 237-238

[29] Suzuki S, Matsunaga K, Ohnishi H, Ishibashi Y. Influences of network delay variation on haptic perception under adaptive reaction force control. In: Proc. the 3rd IEEE Global Conference on Consumer Electronics (GCCE), Tokyo, Japan. 2014. pp. 669-673

[30] Abe T, Komatsu Y, Onishi H, Ishibashi Y. QoE assessment of adaptive

viscoelasticity control in remote control system with haptic and visual senses. In: Proc. IEEE International Conference on Consumer Electronics - Taiwan (ICCE-TW), Chinese Tapei. 2018. pp. 133-134

[31] Ishibashi Y, Taguchi E, Huang P, Tateiwa Y. Robot position control with force information in cooperation between remote robot systems. In: Proc. the 5th International Conference on Control, Automation and Robotics (ICCAR), Beijing, China. 2019. pp. 147-151

[32] Kato H, Ishibashi Y, Ohnishi H, Huang P. Force Adjustment Control for Carrying Object between Remote Robot Systems with Force Feedback. Fukui, Japan: IEICE Technical Report, CQ2022–13. 2022 (In Japanese)

[33] Rikiishi T, Ishibashi Y, Huang P, Miyoshi T, Ohnishi H, Tateiwa Y, et al. Effect of Stabilization Control by Viscocity in Remote Robot System. Chiba, Japan: IEICE Technical Report, MVE2017–19. 2017 (In Japanese)

[34] Arima R, Huang P, Ishibashi Y, Tateiwa Y. Softness Assessment of Objects in Remote Robot System with Haptics: Comparison between Reaction Force Control upon Hitting and Stabilization Control. Tokyo, Japan: IEICE Technical Report, CQ2017–98. 2018 (In Japanese)

[35] Ishibashi Y, Nagasaka M. An adaptive QoS control scheme of avatars in distributed virtual environments. The Journal of The Institute of Image Information and Television Engineers. 2006;**60**(11):1835-1839

[36] Wang X, Huang P, Ishibashi Y, Okuda T, Watanabe H. Influence of network delay on QoS control using neural network in remote robot systems with force feedback. In: Proc. 2020 the 9th International Conference on Networks, Communication and Computing (ICNCC), Tokyo, Japan. 2020

[37] Hashem IAT, Yaqoob I, Anuar NB, Mokhtar S, Gani A, Khan SU. The rise of 'big data' on cloud computing: Review and open research issues. Information Systems. 2015;**47**:98-115

[38] Memos VA, Psannis KE, Ishibashi Y, Kim BG, Gupta BB. An efficient algorithm for media-based surveillance system (EAMSuS) in IoT smart city framework. Future Generation Computer System. 2018;**83**:619-628

[39] Ramesh A, Kambhampati C, Monson J, Drew P. Artificial intelligence in medicine. Annals of the Royal College of Surgeons of England. 2004;**86**:334-338

Chapter 5

Perspective Chapter: Multi-Contact Humanoid Stability for Increased Interaction in Unstructured Environments

Parastoo Dastangoo and Alejandro Ramirez-Serrano

Abstract

This paper presents a practical solution to the problem of multi-legged robot stability moving on unstructured 3D terrains using a multi-contact approach. The solution is based on a modified version of the contact wrench set method, which has been reformulated to utilize three arbitrary contact points representing the physics of contact between the robot and the environment. The new formulations are then used to test the stability of a life-size humanoid robot. The proposed method extends available formulations making the CWS tool suitable for a variety of terrains having various physical and geometrical characteristics including heterogenous coplanar (e.g., stairs and sloped terrain) and noncoplanar surfaces (e.g., gravel, sand, compliant terrains, etc.). The results provided by the new formulations are visualized using a graphical visualization tool, showing the adaptability of the proposed formulation in multi-contact locomotion. The results show that the proposed approach effectively quantifies the robot's stability on a wide range of surfaces and environments, such as mines, industrial facilities, and urban search and rescue operations. Experimental tests on a life-size humanoid are also presented.

Keywords: multi-legged robotics, humanoid stability, multi-contact locomotion, CWS method

1. Introduction

According to the International Labor Organization, 2.3 million work-related deaths happen around the world annually. Work-related incidents also contribute to thousands of disabling work-related injuries every year. These statistics are even more pronounced in work-related tasks that are performed in highly unstructured environments, such as mining and forestry.

A viable approach to keep humans out of harm's way is to provide them with robotic artifacts that can assist them when the associated tasks are performed in high-risk structured/unstructured chaotic and confined environments. In particular, multi-legged artifacts, such as quadrupeds, humanoids, and hybrid robots, are promising devices as they have (at least in theory) the capabilities to penetrate and perform tasks in complex unstructured (potentially compliant environments. However, maintaining the stability of such robots in the mentioned environments is complex and improved mechanisms need to be developeded before multi-legged artificial systems can be effectively deployed in such spaces.

Among the large body of stability mechanisms developed targeting robotic systems, the zero moment point (ZMP) method [1] is one of the most popular. However, ZMP is only able to assess the stability of walking robotic systems moving on flat horizontal surfaces. It cannot be applied to assess stability in terrains having noncoplanar surfaces, unstructured terrains, and deformable (compliant) terrains. The reason is that ZMP cannot determine whether any of the robot's feet slip or not, and it cannot be applied on uneven terrain where foot contacts take place on noncoplanar surfaces. To overcome these aspects, diverse formulations have been proposed capable of capturing the entire robot's dynamics, including friction constraints [2]. Among such formulations, the contact wrench set (CWS) method is a formulation capable of measuring the stability of multi-legged walking robots when the system has multiple noncoplanar contacts where friction cone constraints hold. For this, the method uses a set of allowable wrenches obtained from the contact forces at each individual contact point [3–5]. For a multi-legged robot to maintain its stability, its total contact wrench, calculated from the robot's motion, should lie within the CWS [3, 4]. The CWS method has been enhanced by investigating the extent to which the method is valid and its application in different operation scenarios [3, 6–10]. For instance, the authors in Ref. [10] developed a linear and angular momentum motion planning method using the robot's prespecified contact locations. The authors in Ref. [9] used pre-specified contact locations and CWS to generate whole-body robot motions. Other authors [e.g., [6, 7]] have considered the CWS at the robot's center of mass (CoM) to generate whole-body motion plans. Such work uses nonlinear optimization within the multi-contact domain to generate walking patterns when climbing stairs while the robot is able to employ handrails for additional support. Furthermore, complementary formulations have been developed to compute the robot's CoM feasible acceleration to generate effective walking patterns [11]. Similarly, researchers have developed time-optimal trajectory generation approaches using fixed whole-body robot postures with prespecified contact locations using optimization approaches [e.g., Ref. [8]]. In other publications [e.g., Ref. [12]], authors have combined the ZMP and the CWS methods to maximize their individual advantages.

Although the published studies provide advances in formulating somewhat effective stability methodologies using a set of discrete contact points, the available methodologies fail to fully resolve and provide a generic methodology for stability on continuous surface contacts where the contacts point between the robot and the ground can be perceived as a surface having infinite contact points (e.g., a continuous sole of the robot's foot in contact with a deformable ground surface, such as sand). To solve this issue, many approaches have used a large finite number of individual contact points and contact forces to capture the physics of the continuous surface [13]. However, such approaches are computationally expensive and cannot be effectively employed in real time or in cases where robots might need to move at high speeds. In an attempt to provide a solution to such a problem, the authors in Ref. [13] proposed the use of a closed-form formulation capable of computing the CWS stability explicitly, without the need to make reference to individual contact points. Although effective, such method, is only applicable to contact areas between the robot and the ground having rectangular shapes (e.g., typical shape for humanoid feet). Such a limitation restricts the application of the formulation to particular foot shapes. Furthermore, the methodology cannot be used in robots having partial contact with the ground (e.g., when partial footholds exist e.g., Ref. [14]). In this manuscript, the scope of the CWS method is extended by proposing a modified formulation that is applicable to robots walking on unstructured terrains and when robots move on surfaces that can be perceived as providing an arbitrary number of contact points that reduce to three contact forces. In such an approach, the stability of the robot can be measured on practically any real-world terrain, where each foot of the robot might only have partial contact with the ground (e.g., on rough and rocky terrains). The proposed approach can thus be employed in realistic contact scenarios while reducing the computation time via the use of fewer contact forces when compared to the traditional CWS method. This, in turn, enables us to employ the solution on arbitrary terrain topographies provided there are at least three contact points between the robot and the environment. Hence, the proposed approach provides opportunities to deploy humanoids and other multi-legged robots in real-world complex environments.

Furthermore, the proposed method described in this paper is also suitable to measure the robustness of the robot's stability via a newly developed margin for acceptable contact wrenches [15]. In this paper, radar charts are developed to visualize the robot's stability condition. Such visualization technique enables roboticists to have the tool to better control/guide multi-legged robots moving on unstructured potentially dynamic terrains encompassing surfaces having various geometrical and physical characteristics.

The rest of this paper is organized as follows: Section 2 describes the proposed methodology. Section 3 provides simulation and experimental results, and conclusions are provided in Section 4.

2. Proposed methodology

Multi-contact stability methods, such as the CWS, rely on sensor data information measured on the contact points. Based on the available information such methods shape the robot's stability condition. In general, contact models have been employed as a key element in determining the stability of any multi-legged robot. In this paper, two types of contact styles are considered as a robot locomotes through a given surface: (*i*) *fixed contact* defined when the robot's foot is in the support phase, and (*ii*) *free contact* (i.e., six unconstrained degrees of freedom—DoF) defined when the foot of interest is in its swing phase. In this work shifting between contact modes (e.g., from fix contact to free contact and vice versa) as well as other types of contact modes, such as sliding contact modes, are not considered. Rather the work focuses on determining the stability of the robot when in a fix contact mode. Such a focus does not, however, restrict the methodology to be applied in locomotion transition mode phases as it is assumed that the robot's locomotion is performed in a quasi-static manner where it can be considered to be in a fixed contact mode. That is, the proposed stability mechanisms simplify the mathematical formulations but do not restrict the robot from locomoting at high speeds.

2.1 Physics of friction

To determine whether a given foot, i, of the robot of interest is in the fixed contact mode the exerted force on the system (foot) is checked [16]. For this, the coulomb friction cone on foot i is employed. Accordingly, the contact mode of a box representing the robot's foot remains fixed as long as the contact force at point i, f_c^i, remains inside the Coulomb friction cone [17]. Therefore, for the contact force to remain inside the friction cone, it needs to satisfy Eq. (1):

$$\begin{aligned} f_c^i \cdot n_i &> 0 \\ |f_c^i \cdot t_i| &\leq \mu \cdot \left(f_c^i \cdot n_i\right) \\ |f_c^i \cdot b_i| &\leq \mu \cdot \left(f_c^i \cdot n_i\right) \end{aligned} \tag{1}$$

where (t_i, b_i, n_i) is the local orthogonal contact frame at the contact point C_i, where t_i, b_i, and n_i denote the two tangential, and the surface normal components, respectively, f_c^i denotes the contact force at the point C_i, and μ denotes the static coefficient of friction.

2.2 Contact of surface areas

The physics of a single contact point described above, however, cannot be employed as defined in Eq. (1). Such formulation must be extended to consider the generic case where two surfaces (in contrast to two points) are in contact with each other where the contact between surfaces is to be considered continuous over the corresponding contact area. Under such conditions, one can consider that there exist infinite contact points between the corresponding surfaces where each (and discrete) contact point has a specific contact force (**Figure 1a**). Under such conditions, previously proposed methods require an infinite amount of time to compute the friction cone for each of contact points comprising the contact surface and find the corresponding fix mode boundaries. Fortunately, the authors in Ref. [13] have proven that under the Coulomb friction assumption, rather than dealing with an infinite number of reaction force vectors found on a continuous surface (needed to calculate the corresponding wrench at the center of the surface), a discrete number of friction cones can be defined at diverse points of the contact surface area adding up to the same wrench (**Figure 1b**). Three contact points used to define a given plane is

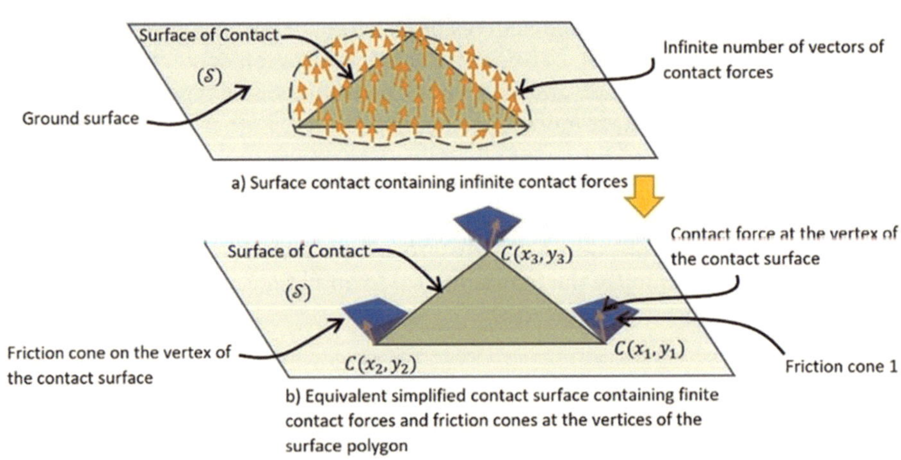

Figure 1.
A surface (e.g., robot's sole) in contact with the ground through surface contact.

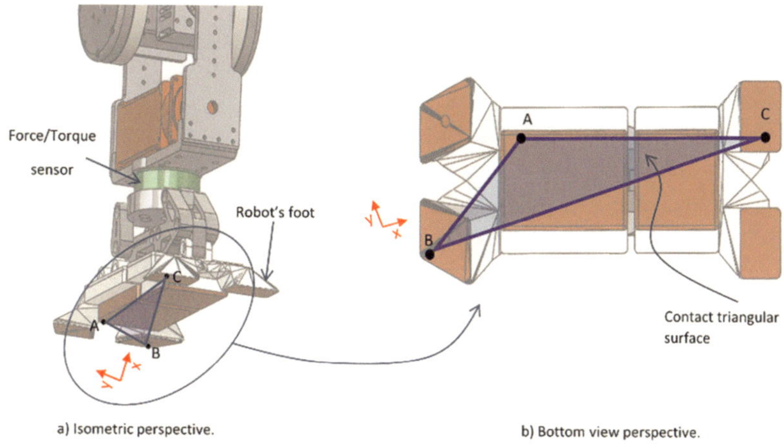

Figure 2.
Contact triangle from isometric and bottom perspectives.

significant when compared to using a large number of contact points, as it provides numerous advantages, including shorter time to compute the given stability, while providing a more generic solution.

Furthermore, a triangular surface is a good, simplified approximation of any typical contact between a robot's foot (or feet) and the ground when the foot (feet) is (are) in its (their) support (*fixed contact*) phase. This is true when the robot's foot is in contact with any rigid surface, especially rugged ones, such as stepping on rocky surfaces, where the foot makes contact only with a small set of discrete contact points. This approximation is also effective and applicable for deformable surfaces (e.g., loose sandy terrains) assuming the terrain deformation takes place over a sufficiently long period of time such that the deformation could be considered quasi-static. In this paper, standing on a single point (e.g., pegs) or in line contacts (e.g., stepping on a long and narrow surface—funambulism) is not considered. **Figure 2** illustrates a set of three random contact points, A, B, and C forming a virtual triangular contact area ABC when a robot's foot lands on an unstructured rigid terrain. Such a surface is considered virtual as it might not fully define the real surface but rather a hypothetical plane where the robot is setting its foot.

2.3 Triangular surfaces and corresponding wrench cone

With the virtual triangular surface area ABC, a novel analytical formulation can be developed for calculating the wrench cone using the known Fourier-Motzkin method [13].

Without loss of the generality, a geometrical frame of reference defining the virtual triangular surface can be produced by aligning the x-axis alongside the BC edge of the triangular surface, with the orthogonal y-axis defined in the same plane. When the robot moves over a flat rigid terrain the virtual triangle can be considered as part of the real contact surface.

Based on the location of the contact points A, B, and C defining the virtual triangle (shown in **Figures 1** and **2**) each having coordinates (x_i, y_i) the following properties are computed (**Figure 3**).

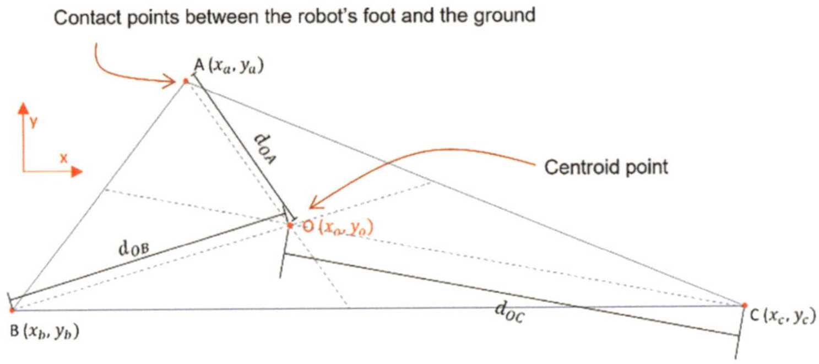

Figure 3.
A general contact triangle and its centroid.

For a given triangle, the coordinates of its centroid (x_O, y_O), as well as the distances, d_{OA}, d_{OB}, and d_{OC} from the triangle's centroid to each of the vertices can be computed per Eqs. (2) and (3), respectively.

$$x_O = \frac{x_A + x_B + x_C}{3};\ y_O = \frac{y_A + y_B + y_C}{3} \tag{2}$$

$$d_{OA} = \begin{pmatrix} dx_{OA} = (x_A - x_O) \\ dy_{OA} = (y_A - x_O) \end{pmatrix};$$

$$d_{OB} = \begin{pmatrix} dx_{OB} = (x_B - x_O) \\ dy_{OB} = (y_B - x_O) \end{pmatrix}; \tag{3}$$

$$d_{OC} = \begin{pmatrix} dx_{OC} = (x_C - x_O) \\ dy_{OC} = (y_C - x_O) \end{pmatrix};$$

$$|dx_{OA}| = x_{OA}, |dx_{OB}| = x_{OB}, |dx_{OC}| = x_{OC}$$

$$|dy_{OA}| = y_{OA}, |dy_{OB}| = y_{OB}, |dy_{OC}| = y_{OC}$$

Under the above formulations, it can be shown that the contact forces on the contact surface are equivalent to a set of forces at the contact vertices comprising the virtual triangular surface. Therefore, in what follows, three contact forces at points A, B, and C and the corresponding torque at the triangle's centroid are used to determine the physical interaction that exists between the foot and the ground. To achieve this, a set of mathematical formulations, Eqs. (4) and (5), developed as part of this research work are used.

Figure 4 illustrates a top-down view of one of the robot's feet in contact with the ground, including the contact forces and contact torques, exerting on the contact points. The complete system of forces and torques is formulated as per Eqs. (4) and (5).

$$\begin{aligned} f_x &= f_x^A + f_x^B + f_x^C \\ f_y &= f_y^A + f_y^B + f_y^C \\ f_z &= f_z^A + f_z^B + f_z^C \end{aligned} \tag{4}$$

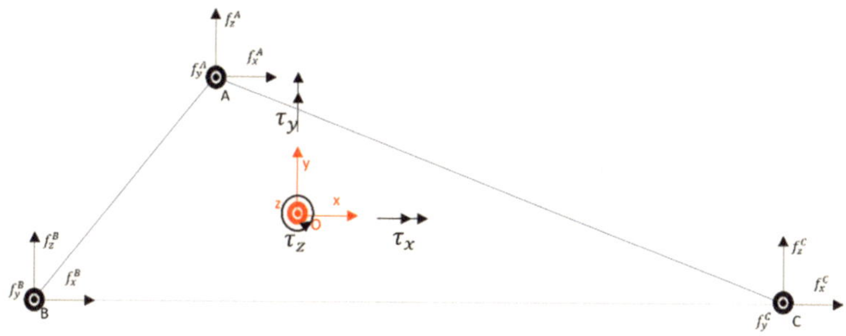

Figure 4.
The location and orientation of forces with respect to the frame of reference located on the triangle's centroid.

$$\begin{aligned} \tau_x &= f_z^1(y_{OA}) - f_z^2(y_{OB}) - f_z^3(y_{OC}) \\ \tau_y &= f_z^1(x_{OA}) + f_z^2(x_{OB}) - f_z^3(x_{OC}) \\ \tau_z &= f_x^3(y_{OC}) + f_x^2(y_{OB}) - f_x^1(y_{OA}) + f_y^3(x_{OC}) - f_y^2(x_{OB}) - f_y^1(x_{OA}) \end{aligned} \tag{5}$$

For this, a theorem is used, where the wrench cone at each of the three contact-point systems can be expressed (Theorem 1).

Theorem 1: A system with three contact points f^A,f^B,f^C obeying the Coulomb friction rules exist, if and only if there exists a contact wrench cone with wrench = $(f_x, f_y, f_z, \tau_x, \tau_y, \tau_z)$ at the centroid, $O(x_0,y_0)$, of the virtual contact triangle satisfying Eq. (6).

$$\begin{aligned} |f_x| &\le \frac{y_{OA}}{y_{sum}} \cdot \mu f_z. \\ |f_y| &\le \frac{x_{OC}}{x_{sum}} \cdot \mu f_z \\ f_z &> 0 \\ -y_{OB} \cdot f_z \le &\ \tau_x \le y_{OA} \cdot f_z \\ -x_{OC} \cdot f_z \le &\ \tau_y \le x_{OB} \cdot f_z \\ \tau_{min} \le &\ \tau_z \le \tau_{max} \end{aligned} \tag{6}$$

where

$$x_{Oi} = |x_i - x_O|, x_O = \frac{x_A + x_B + x_C}{3}, i = A, B, C$$

$$y_{Oi} = |y_i - y_O|, y_O = \frac{y_A + y_B + y_C}{3}, i = A, B, C$$

$$x_{sum} = x_{OA} + x_{OB} + x_{OC}$$

$$y_{sum} = y_{OA} + y_{OB} + y_{OC}$$

$$\tau_{min} = -\mu(x_{OC} + y_{OA})f_z + |y_{sum} \cdot f_x - \mu\tau_x| + |x_{sum} \cdot f_y - \mu\tau_y|$$

$$\tau_{max} = \mu(x_{OC} + y_{OA})f_z - |y_{sum} \cdot f_x + \mu\tau_x| - |x_{sum} \cdot f_y + \mu\tau_y|$$

Theorem 1 guarantees that if the CWC obeys Eq. (6), there exists a set of three contact forces that fulfill Eq. (1) (i.e., satisfy the Coulomb friction rules). The significance of this theorem is twofold:

i. Theorem 1 connects the physics of continuous surface contact to the physics of a wrench cone in the centroid of the surface of interest.

ii. It simplifies the determination of the robot's stability condition by examining the wrench just at the centroid of the surface rather than the need to verify the entire surface.

Therefore, Theorem 1 provides the mechanism to demonstrate whether a foot or robot of interest is stable or not via the value of the wrench at the centroid of a continuous (real or virtual) contact surface. The added benefit of the proposed method is that it is suitable for implementation on any real robot which has a force/torque (F/T) sensor positioned at the interface between the robot and the contact with the surface (e.g., ankle of the robot's foot or at any extremity e.g., arms) that the robot might use to locomote in any style (e.g., crawling). The captured F/T sensor data can then be used to compute the corresponding wrench at the centroid of the corresponding contact triangular surface.

Eq. (6) determines the stability condition of multi-legged robots in binary logical terms (i.e., $1 \rightarrow$ stable; $0 \rightarrow$ unstable). If all six inequalities in Eq. (6) are satisfied the robot is said to be stable, otherwise, it is unstable.

However, to understand how stable a stable robot is a stability margin is needed as well as a visualization tool, where users can determine the conditions in which the robot is stable and what movements the robot might need to avoid which might cause it to approach instability.

Proof:

In order to prove that the previous statements are satisfied, there is a need to relate Eqs. (4) and (5) to the Coulomb friction law, so that they can be expressed as positive combinations enabling the use of linear algebra algorithms. In order to achieve this, all force and torque variables (*e.g.*, $f^i_x, f^i_y, f^i_z f_x, f_y, f_z, \tau_x, \tau_y, \tau_z$) at each contact point (between the robot and the environment) and the centroid of the triangle formed by three contact points are normalized by defining a set of new variables $(k_1, k_2, k_3, c_1, c_2, \alpha^i_x, \alpha^i_y, \beta^i_z)$ according to the Coulomb friction law. It is worth mentioning that in such a procedure there exist 15 variables $f^i_x, f^i_y, f^i_z f_x, f_y, f_z, \tau_x, \tau_y, \tau_z$ $(i = 1, 2, 3)$, which are substituted with 14 new variables $k_1, k_2, k_3, c_1, c_2, \alpha^i_x, \alpha^i_y, \beta^i_z$ $(i = 1, 2, 3)$. Therefore, while normalizing the variables, one variable is reduced as the first step of the Fourier-Motzkin elimination method [18].

In order to normalize the forces and torques while satisfying Eq. (1), there is a need to define a new set of variables as per Eq. (7).

$$k_1 := \frac{f_x}{\mu f_z};\ k_2 := \frac{f_y}{\mu f_z};\ k_3 := \frac{\tau_z}{\mu(X+Y)f_z} \qquad (7)$$

$$c_1 := \frac{\tau_x}{y_{sum} f_z};\ c_2 := \frac{\tau_y}{x_{sum} f_z}$$

Normalizing the contact forces employing the parameters α^i_x, α^i_y, and β^i_z produce the expressions in Eq. (8):

$$\alpha_x^i := \frac{f_x^i}{\mu f_z^i};\alpha_y^i := \frac{f_y^i}{\mu f_z^i};\beta_z^i := \frac{f_z^i}{f_z} \tag{8}$$

Recall:

$$\sum_i f_z^i = f_z \textit{ with } \forall i \in N : f_z^i > 0$$

It must be noted that Eq. (8) produces nine expressions ($i = 1 \textit{ to } 3$), where i represents one of the three contact points. To obtain a set of suitable expressions to enable developing a solution for the Fourier-Motzkin elimination procedure, the 11 parameters used in Eq. (9) are defined as:

$$\begin{gathered} y_{sum} := y_{OA} + y_{OB} + y_{OC} \\ x_{sum} := x_{OA} + x_{OB} + x_{OC} \\ Y_{OA} := \frac{y_{OA}}{y_{sum}};Y_{OB} := \frac{y_{OB}}{y_{sum}};Y_{OC} := \frac{y_{OC}}{y_{sum}} \\ X_{OA} := \frac{x_{OA}}{x_{sum}};X_{OB} := \frac{x_{OB}}{x_{sum}};X_{OC} := \frac{x_{OC}}{x_{sum}} \\ X := \max(X_{OA}, X_{OB}, X_{OC}) \\ Y := \max(Y_{OA}, Y_{OB}, Y_{OC}) \\ p_x = \frac{x_{sum}}{x_{sum} + y_{sum}};p_y = \frac{y_{sum}}{x_{sum} + y_{sum}} \\ Q := X.p_x + Y.p_y \end{gathered} \tag{9}$$

By considering Eqs. (4) and (5), as well as the new set of variables defined in Eqs. (7) and (8) lead to Eqs. (10):

$$1 = \beta_z^1 + \beta_z^2 + \beta_z^3 \tag{10a}$$

$$c_1 = \beta_z^1.Y_{OA} - \beta_z^2.Y_{OB} - \beta_z^3.Y_{OC} \tag{10b}$$

$$c_2 = \beta_z^1.X_{OA} + \beta_z^2.X_{OB} - \beta_z^3.X_{OC} \tag{10c}$$

$$p_x = \frac{x_{sum}}{x_{sum} + y_{sum}};p_y = \frac{y_{sum}}{x_{sum} + y_{sum}} \tag{10d}$$

$$k_3 = p_x\left(-\alpha_y^1.\beta_z^1.X_{OA} - \alpha_y^2.\beta_z^2.X_{OB} + \alpha_y^3.\beta_z^3.X_{OC}\right) + p_y\left(-\alpha_x^1.\beta_z^1.Y_{OA} + \alpha_x^2.\beta_z^2.Y_{OB} + \alpha_x^3.\beta_z^3.Y_{OC}\right) \tag{10e}$$

$$\forall i, |\alpha_x^i| \leq 1 \tag{10f}$$

$$\forall i, |\alpha_y^i| \leq 1 \tag{10g}$$

$$\forall i, \beta_z^i > 0 \tag{10h}$$

Eqs. (10) represents a set of eight inequalities based on variables $k_1, k_2, k_3, c_1, c_2, \alpha_x^i, \alpha_y^i$, and β_z^i. In order to apply the Minkowski sum to the set of inequalities in Eqs. (10), the variables $\gamma_x, \gamma_y, \gamma_x', \gamma_y'$ are defined as per Eq. (11).

$$\begin{aligned}\gamma_x &= \alpha_x^2.\beta_z^2.Y_{OB} + \alpha_x^3.\beta_z^3.Y_{OC}\\ \gamma_y &= \alpha_y^1.\beta_z^1.X_{OA} + \alpha_y^2.\beta_z^2.X_{OB}\\ \gamma_x' &= \alpha_x^1.\beta_z^1.Y_{OA}\\ \gamma_y' &= \alpha_y^3.\beta_z^3.X_{OC}\end{aligned} \tag{11}$$

The variables defined in Eq. (11) can then be used to derive a simpler system of inequalities based on the properties of α_x^i, α_y^i and β_i^z. From Eqs. (8), (10), and (11), it is concluded that $\sum_i \beta_z^i = 1$ and $\beta_i^z > 0$. Furthermore, from Eqs. (1) and (8), it is clear that the parameter α (defined in Eq. (8)) lies within the range of $[-1, 1]$. By applying these properties in the expressions in Eq. (11), the following set of inequalities is obtained (Eqs. (12)).

$$\begin{aligned}|\gamma_x| &\le Y_{OB}.\beta_z^2 + Y_{OC}.\beta_z^3\\ |\gamma_x'| \le Y_{OA}.\beta_z^1 &\Rightarrow |\gamma_x'| - |\gamma_x| \le c_1\\ |\gamma_x'| + |\gamma_x| &\le \max(Y_{OA}, Y_{OB}, Y_{OC}) = Y\end{aligned} \tag{12a}$$

and,

$$\begin{aligned}|\gamma_y| &\le X_{OA}.\beta_z^1 + X_{OB}.\beta_z^2\\ |\gamma_x'| \le X_{OC}.\beta_z^3 &\Rightarrow |\gamma_y| - |\gamma_y'| \le c_2\\ |\gamma_y'| + |\gamma_y| &\le \max(X_{OA}, X_{OB}, X_{OC}) = X\end{aligned} \tag{12b}$$

Given the fact that the Fourier-Motzkin method algorithm works based on eliminating variables in each step, to proceed with this algorithm on the inequalities in Eqs. (12), some of the variables should be eliminated. Therefore, the following lemma describes the procedure for removing β_z^i by finding the boundaries of $|\gamma_x|$, $|\gamma_x'|$, $|\gamma_y|$, and $|\gamma_y'|$ based on Eqs. (12).

Lemma 1) *Following boundaries are derived based on* Eqs. (12):

$$2|\gamma_x'| \le Y + c_1 \tag{13a}$$

$$2|\gamma_x| \le Y - c_1 \tag{13b}$$

$$2|\gamma_y'| \le X - c_2 \tag{13c}$$

$$2|\gamma_y| \le X + c_2 \tag{13d}$$

Proof of lemma 1:

Summing up the both sides of inequalities $|\gamma_x'| - |\gamma_x| \le c_1$ and $|\gamma_x'| + |\gamma_x| \le Y$ (refer to Eq. (12a)) results into $2|\gamma_x'| \le Y + c_1$, which is Eq. (13a); therefore, Eq. (13a) is proved. By similar calculations on Eq. (12b), Eq. (13d) can also be proved (summing both sides of inequalities). To prove Eq. (13b), it is required to recall the definition of Y (Eq. (9)):

Recall:

$$Y = \max(Y_{OA}, Y_{OB}, Y_{OC})$$

Therefore,

$$\begin{aligned} Y_{OB} &\le Y \\ \Rightarrow Y_{OB} + Y_{OB} &\le Y + Y_{OB} \\ \Rightarrow 2Y_{OB} &\le Y + Y_{OB} \\ \Rightarrow 2Y_{OB}\beta_z^2 &\le (Y + Y_{OB})\beta_z^2 \end{aligned} \tag{14}$$

Similarly,

$$2Y_{OC}\beta_z^3 \le (Y + Y_{Oc})\beta_z^3 \tag{15}$$

Now, in a similar procedure, by summing both sides of inequalities (14) and (15), *the following inequality is obtained:*

$$2\left(Y_{OB}\beta_z^2 + Y_{OC}\beta_z^3\right) \le (Y + Y_{OB})\beta_z^2 + (Y + Y_{Oc})\beta_z^3 \tag{16}$$

Again, definition of Y (Eq. (9)) *leads to:*

$$\begin{aligned} Y_{OA} &\le Y \\ \Rightarrow 0 &\le Y - Y_{OA} \\ \Rightarrow 0 &\le (Y - Y_{OA})\beta_z^1 \end{aligned}$$

The following inequality is derived by adding both sides of inequalities (15) and (16)*:*

$$\begin{aligned} 2\left(Y_{OB}\beta_z^2 + Y_{OC}\beta_z^3\right) &\le (Y - Y_{OA})\beta_z^1 + (Y + Y_{OB})\beta_z^2 + (Y + Y_{Oc})\beta_z^3 \\ \Rightarrow 2\left(Y_{OB}\beta_z^2 + Y_{OC}\beta_z^3\right) &\le Y\left(\beta_z^1 + \beta_z^2 + \beta_z^3\right) - \left(\beta_z^1 Y_{OB} - \beta_z^2 Y_{OB} - \beta_z^3 Y_{OC}\right) \end{aligned} \tag{17}$$

By substituting $\beta_z^1 + \beta_z^2 + \beta_z^3 = 1$ *(refer to* Eq. (10a)*) and* $\beta_z^1 Y_{OB} - \beta_z^2 Y_{OB} - \beta_z^3 Y_{OC} = c_1$ *(refer to* Eq. (10b)*), into the above inequality* (Eq. (17)) *the inequality in* Eq. (8) *is obtained:*

$$2\left(Y_{OB}\beta_z^2 + Y_{OC}\beta_z^3\right) \le Y - c_1 \tag{18}$$

Recall from Eq. (12a):

$$\begin{aligned} |\gamma_x| &\le Y_{OB}.\beta_z^2 + Y_{OC}.\beta_z^3 \\ \Rightarrow 2|\gamma_x| &\le 2\left(Y_{OB}\beta_z^2 + Y_{OC}\beta_z^3\right) \end{aligned} \tag{19}$$

Inequalities (18) and (19) *conclude that* $2|\gamma_x| \le Y - c_1$*, which means that* Eq. (13b) *is proved. By implementing a similar procedure on* Eq. (12b), Eq. (13c) *can be proved as well. Therefore, the proof of lemma 1 is completed.* ∎

The proposed method outlined in the paper uses a reduced number of variables when compared to other methods (e.g., CWS). However, to solve for the robot's stability, it is required to reduce the number of variables. For this, the Fourier-Motzkin elimination method is used to reduce the number of variables by one at each step of the process. That is, the β_z^i variables are eliminated through the procedure outlined in Eqs. (12) to (13) (refer to lemma 1). Subsequently, the proposed variable elimination method c_1 and c_2 are eliminated from the set of inequalities in (13). For this, the upper and lower bounds of c_1 and c_2 are determined based on Eq. (21). Such method is heavily based on the geometrical features of the virtual support triangular area depicted in **Figure 4**.

Without loss of generality, the vertices of the defined virtual contact triangle can be named in such a way that the x coordinates of the vertices satisfy Eqs. (20).

$$x_{OC} \geq x_{OB} \geq x_{OA} \tag{20a}$$

and,

$$y_{OA} \geq (y_{OB} = y_{OC}) \tag{20b}$$

Figure 5 illustrates the possible different contact triangles where Eqs. (20) is satisfied (thus, any three contact points can be used in the proposed approach provided the X-axis of the contact triangle is defined to be aligned with the BC line.

Eqs. (20) results from the fact that the distances $x_{OC}, x_{OB}, x_{OA}, y_{OA}, y_{OB},$ and y_{OC} are greater than or equal to zero values (refer to Eq. (3)). By substituting Eq. (10a) in Eqs. (10b) and (10c), the boundary values of c_1 and c_2 (Eqs. (10)) are computed as:

$$c_1 \in [\min(c_1),\ \max(c_1)]$$

$$c_2 \in [\min(c_2),\ \max(c_2)]$$

where

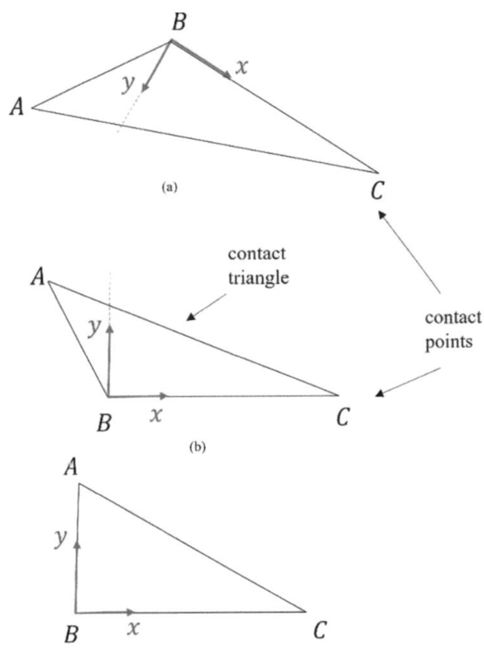

Figure 5.
Naming the vertices of the contact triangle to meet Eqs. (21a) and (21b).

$$\max(c_1) = \max\left(\beta_z^1.Y_{OA} - \beta_z^2.Y_{OB} - \beta_z^3.Y_{OC}\right) = 1 \times Y_{OA} + 0 + 0 = Y_{OA}$$

$$\min(c_1) = \min\left(\beta_z^1.Y_{OA} - \beta_z^2.Y_{OB} - \beta_z^3.Y_{OC}\right) = 0 + 0 - 1 \times Y_{OB} = -Y_{OB}$$

$$\max(c_2) = \max\left(\beta_z^1.X_{OA} + \beta_z^2.X_{OB} - \beta_z^3.X_{OC}\right) = 0 + 1 \times X_{OB} + 0 = X_{OB}$$

$$\min(c_2) = \min\left(\beta_z^1.X_{OA} + \beta_z^2.X_{OB} - \beta_z^3.X_{OC}\right) = 0 + 0 - 1 \times X_{OC} = -X_{OC}$$

The above formulations give rise to Eq. (21).

$$\begin{aligned} -Y_{OB} \le c_1 \le Y_{OA} \\ -X_{OC} \le c_2 \le X_{OB} \end{aligned} \tag{21}$$

Accordingly, the coordinates for Y and X (defined in Eq. (9)) can be denoted as Y_{OA} and X_{OC}, respectively, as per the following expression:

$$X = \max(x_{OA}, x_{OB}, x_{OC}) = X_{OC}$$

$$Y = \max\left(y_{OA}, y_{OB}, y_{OC}\right) = Y_{OA}$$

The result of the above process provides a set of inequalities that describe the boundaries of the defined variables $(\gamma_x', \gamma_x, k_1, k_2, c_1, \text{and} c_2)$ as per Eq. (22):

$$\begin{aligned} k_1 &= \gamma_x + \gamma_x' \\ k_2 &= \gamma_y + \gamma_y' \\ k_3 &= p_x\left(\gamma_y' - \gamma_y\right) + (\gamma_x - \gamma_x') \\ 2|\gamma_x'| &\le Y_{OA} + c_1; 2|\gamma_x| \le Y_{OA} - c_1 \\ 2|\gamma_y'| &\le X_{OC} - c_2;\ 2|\gamma_y| \le X_{OC} + c_2 \\ -Y_{OB} &\le c_1 \le Y_{OA} \\ -X_{OC} &\le c_2 \le X_{OB} \end{aligned} \tag{22}$$

Eq. (22) describes all $\gamma's$ with respect to γ_x, however, there is a need to eliminate the $\gamma's$ parameters. Such elimination is achieved via the fourth step of the Fourier-Motzkin procedure as described below via Eqs. (23) and (24), where all inequalities are in terms of $2p_y\gamma_x$.

$$2p_y\gamma_x \le p_y(Y_{OA} - c_1) \tag{23a}$$

$$2p_y\gamma_x \le p_y(Y_{OA} + c_1) + 2p_y.k_1 \tag{23b}$$

$$2p_y\gamma_x \le p_x(X_{OC} + c_2) + k_3 + p_y.k_1 - p_x.k_2 \tag{23c}$$

$$2p_y\gamma_x \le p_x(X_{OC} - c_2) + k_3 + p_y.k_1 + p_x.k_2 \tag{23d}$$

$$2p_y\gamma_x \ge -p_y(Y_{OA} - c_1) \tag{24a}$$

$$2p_y\gamma_x \ge -p_y(Y_{OA} + c_1) + 2p_y.k_1 \tag{24b}$$

$$2p_y\gamma_x \geq -p_x(X_{OC} + c_2) + k_3 + p_y.k_1 - p_x.k_2 \tag{24c}$$

$$2p_y\gamma_x \geq -p_x(X_{OC} - c_2) + k_3 + p_y.k_1 + p_x.k_2 \tag{24d}$$

Since the right-hand side terms of Eqs. (24) have higher values when compared to the left-hand side terms, all $2p_y\gamma_x$ terms in Eqs. (23) and (24) can be eliminated, resulting in the following set of 16 inequalities.

$$24a \leq 23a \rightarrow c_1 \leq Y_{OA}$$

$$24a \leq 23b \rightarrow k_1 \geq -Y_{OA}$$

$$24a \leq 23c \rightarrow -k_3 - p_y.k_1 + p_x.k_2 + p_y.c_1 - p_x.c_2 \leq X_{OC}.p_x + Y_{OA}.p_y$$

$$24a \leq 23d \rightarrow -k_3 - p_y.k_1 - p_x.k_2 + p_y.c_1 + p_x.c_2 \leq X_{OC}.p_x + Y_{OA}.p_y$$

$$24b \leq 23a \rightarrow k_1 \leq Y_{OA}$$

$$24b \leq 23b \rightarrow c_1 \geq -Y_{OA}$$

$$24b \leq 23c \rightarrow -k_3 + p_y.k_1 + p_x.k_2 - p_y.c_1 - p_x.c_2 \leq X_{OC}.p_x + Y_{OA}.p_y$$

$$24b \leq 23d \rightarrow -k_3 - p_y.k_1 - p_x.k_2 + p_y.c_1 + p_x.c_2 \leq X_{OC}.p_x + Y_{OA}.p_y$$

$$24c \leq 23a \rightarrow k_3 + p_y.k_1 - p_x.k_2 + p_y.c_1 - p_x.c_2 \leq X_{OC}.p_x + Y_{OA}.p_y$$

$$24c \leq 23b \rightarrow k_3 - p_y.k_1 - p_x.k_2 - p_y.c_1 - p_x.c_2 \leq X_{OC}.p_x + Y_{OA}.p_y$$

$$24c \leq 23c \rightarrow c_2 \geq -X_{OC}$$

$$24c \leq 23d \rightarrow k_2 \geq -X_{OC}$$

$$24d \leq 23a \rightarrow k_3 + p_y.k_1 + p_x.k_2 + p_y.c_1 + p_x.c_2 \leq X_{OC}.p_x + Y_{OA}.p_y$$

$$24d \leq 23b \rightarrow k_3 - p_y.k_1 + p_x.k_2 - p_y.c_1 + p_x.c_2 \leq X_{OC}.p_x + Y_{OA}.p_y$$

$$24d \leq 23c \rightarrow k_2 \leq X_{OC}$$

$$24d \leq 23d \rightarrow c_2 \leq X_{OC}$$

The $\rightarrow$ symbol in the previous set of 16 inequalities is used to denote that the left-hand side of the corresponding inequality results in the right-hand inequality. Substituting the expression for Q in Eq. (9), $Q = X_{OC}.p_x + Y_{OA}.p_y$, into the above 16 inequalities result in Eqs. (25) and (26).

$$k_3 \leq Q - p_y.k_1 + p_x.k_2 - p_y.c_1 + p_x.c_2$$

$$k_3 \leq Q + p_y.k_1 + p_x.k_2 + p_y.c_1 + p_x.c_2$$

$$k_3 \leq Q - p_y.k_1 - p_x.k_2 - p_y.c_1 - p_x.c_2$$

$$k_3 \leq Q + p_y.k_1 - p_x.k_2 + p_y.c_1 - p_x.c_2$$

Therefore,

$$k_3 < Q - p_y|k_1 + c_1| - p_x|k_2 + c_2| \tag{25}$$

Similarly,

$$
\begin{aligned}
k_3 &\geq -Q - p_y.k_1 + p_x.k_2 + p_y.c_1 - p_x.c_2 \\
k_3 &\geq -Q - p_y.k_1 - p_x.k_2 + p_y.c_1 + p_x.c_2 \\
k_3 &\geq -Q + p_y.k_1 + p_x.k_2 - p_y.c_1 - p_x.c_2 \\
k_3 &\geq -Q + p_y.k_1 - p_x.k_2 - p_y.c_1 + p_x.c_2
\end{aligned}
$$

Therefore,

$$k_3 \geq -Q + p_y|k_1 - c_1| + p_x|k_2 - c_2| \tag{26}$$

and,

$$
\begin{aligned}
&k_1 \in [-Y_{OA}, +Y_{OA}],\ k_2 \in [-X_{OC}, +X_{OC}] \\
&c_1 \in [-Y_{OB}, +Y_{OA}],\ c_2 \in [-X_{OC}, +X_{OB}]
\end{aligned}
$$

Now, by eliminating variable k_3 from the set of inequalities, the Fourier-Motzkin method can continue. As discussed, eliminating variables step by step enables to determine the boundaries of all variables. Eliminating the last implicit variable k_3 from Eqs. (25) and (26) results in Eq. (27).

$$
\begin{gathered}
Q - p_y|k_1 + c_1| - p_x|k_2 + c_2| \geq -Q + p_y|k_1 - c_1| + p_x|k_2 - c_2| \\
p_y(|k_1 - c_1| + |k_1 + c_1|) + p_x(|k_2 + c_2| + |k_2 - c_2|) \leq 2Q \\
p_y(\max(|k_1|, |c_1|) + p_x(\max(|k_2|, |c_2|) \leq Q \\
X_{OC}.p_x + Y_{OA}.p_y \leq Q
\end{gathered} \tag{27}
$$

Following the definition of Q (Eq. (9)), Eq. (27) is always valid. The result of the above process is that the variables of interest (i.e., k_1, k_2, k_3, c_1, c_2,α^i_x, α^i_y, β^i_z, and β_z) are real values where each of such variables is obtained by changing the corresponding defined variable to the initial variables (i.e., f^i_x,f^i_y,f^i_z,f_x,f_y, τ_x, τ_y and τ_z). This process leads to the following set of expressions (Eq. (28)), which proves Theorem 1:

$$
\begin{gathered}
|f_x| \leq \frac{y_{OA}}{y_{sum}}.\mu f_z \\
|f_y| \leq \frac{x_{OC}}{x_{sum}}.\mu f_z \\
f_z > 0 \\
-y_{OB}.f_z \leq \tau_x \leq y_{OA}.f_z \\
-x_{OC}.f_z \leq \tau_y \leq x_{OB}.f_z \\
\tau_{min} \leq \tau_z \leq \tau_{max}
\end{gathered} \tag{28}
$$

where

$$x_{Oi} = |x_i - x_O|, x_O = \frac{x_A + x_B + x_C}{3},$$

$$i = A,B,C$$

$$y_{Oi} = |y_i - y_O|, y_O = \frac{y_A + y_B + y_C}{3},$$

$$i = A,B,C$$

$$x_{sum} = x_{OA} + x_{OB} + x_{OC}$$

$$y_{sum} = y_{OA} + y_{OB} + y_{OC}$$

$$\tau_{min} = -\mu(x_{OC} + y_{OA})f_z + |y_{sum} f_x - \mu\tau_x| + |x_{sum} f_y - \mu\tau_y|$$

$$\tau_{max} = \mu(x_{OC} + y_{OA})f_z - |y_{sum} f_x + \mu\tau_x| - |x_{sum} f_y + \mu\tau_y| \quad \blacksquare$$

Eq. (28) provides the stability condition of the corresponding robot's contact with the environment. That is, if the forces and torque values read from the corresponding force and torque (F/T) sensor on the robot (e.g., ankle) satisfy Eq. (28), then, such conditions indicate that the robot is stable. Otherwise, the robot is unstable. The stability results provided from Eq. (28), however, is a binary value (stable or not stable), which does not provide the degree of stability with which the given robot might be (How stable is the robot stable?). Furthermore, such results do not provide information to help guide the motions of the robot such that it does not move toward an unstable state or becomes marginally stable.

2.4 Stability margin and visualizing stability condition

Now that a set of mathematical formulations, Eq. (28), has been developed to enable to determine the stability condition of legged bipedal robots in binary logical terms (i.e., 1 → stable; 0 → unstable), a stability margin is needed. A stability margin based on the parameters used in this work would enable us to know how stable the robot is or how close the robot is from becoming unstable. In this work, a stability margin based on the five parameters used in the proposed formulations (i.e., f_x, f_y, f_z, τ_x, τ_y, and τ_z) is provided to enable us to know how stable the robot is or how close the robot is from becoming unstable. To this aim, a radar chart is proposed as a practical solution to visualize multivariate data on a 2-dimensional surface. Each variable of interest has an axis starting from a common center point from where the axes for the remaining variables emanate. On each axis, the acceptable range of each variable is represented. The maximum values of the five parameters form a polygon called *stability polygon*. While the values of the five parameters computed based on the state of the robot are used to form another polygon, which is referred to in this paper as the *data polygon* in **Figure 6**. As long as the *data polygon* lies inside the stability polygon, as shown in **Figure 6a**, all inequalities are satisfied, and the robot is said to be stable. In contrast, if one or more of the vertices of the *data polygon* crosses the boundaries of the *stability polygon* the robot becomes unstable as illustrated in **Figure 6b**.

The proposed radar graph is divided into different fuzzy regions each indicating how stable the robot is. Each region can be defined based on user preferences, desired levels of desired stability, or robot characteristics. To illustrate such aspects,, in this paper, the allowable range for each of the f_x, f_y, f_z, τ_x, τ_y, and τ_z variables are divided into five severity regions. Such fuzzy stability characteristics are then applied to a life-size humanoid robot having 29 DoF. These stability levels are expressed as fuzzy variables

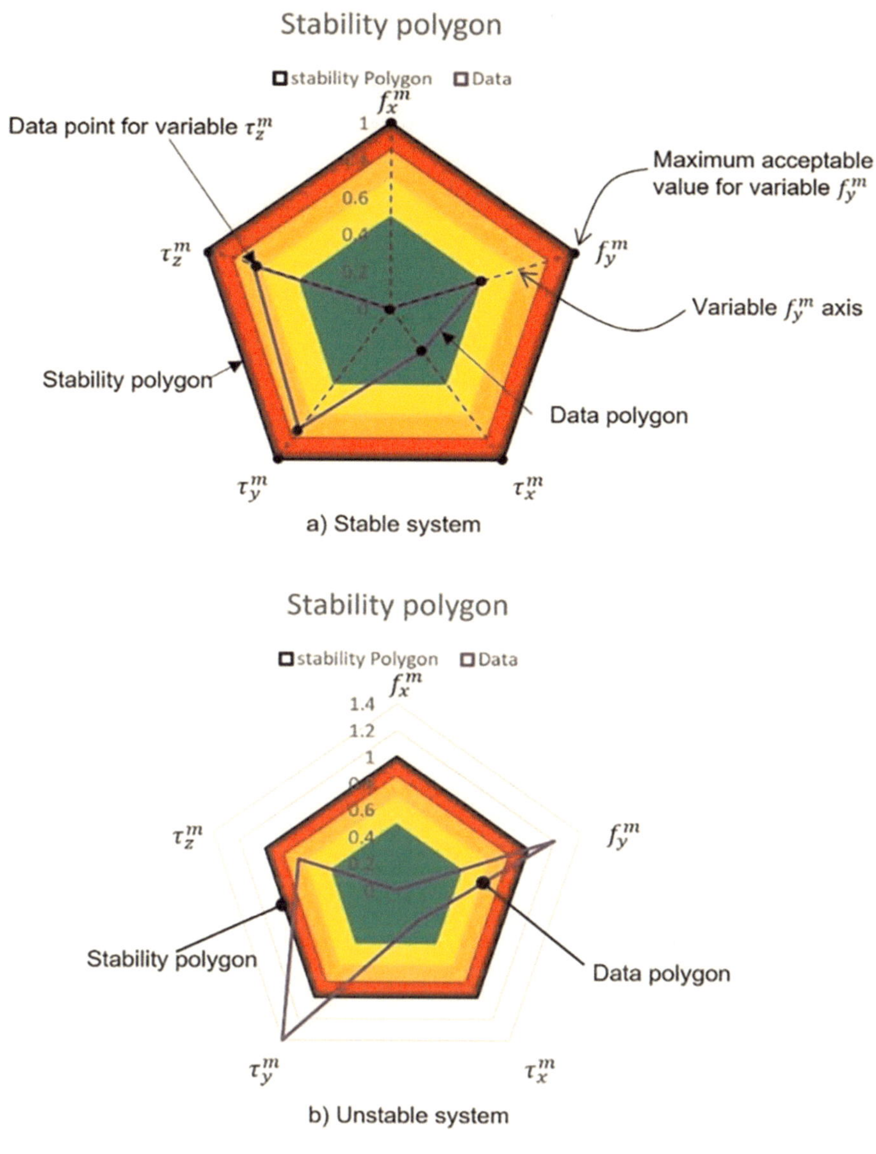

Figure 6.
Stability polygon on a radar graph.

and their ranges are presented in **Table 1**. If the values of the five parameters are inside the region closer to the center of the radar graph, it indicates that the robot is more stable when compared to a case where the parameter values are in regions closer to the borders of polygons, which would indicate that the robot is approaching instability.

To build the radar graph the five parameters f_x, f_y, τ_x, τ_y, and τ_z are normalized based on the six variables f_x, f_y, f_z, τ_x, τ_y, and τ_z to obtain f_x^m, f_y^m, τ_x^m, τ_y^m, and τ_z^m. Among all the six variables, f_z has a direct impact on each of the five normalized variables and as a result does not appear as a normalized variable itself. The formulations to obtain f_x^m, f_y^m, τ_x^m, τ_y^m, τ_z^m are presented in Eq. (29).

Color	Range	Severity
Green	[0 to 0.5]	Highly stable
Yellow	(0.5 to 0.7]	Stable
Orange	(0.7 to 0.85]	Moderate-risk stability
Red	(0.85 to 0.95]	High-risk stability
Dark red	(0.95 to 1]	Approaching instability

Table 1.
Stability regions in the radar chart.

$$
\begin{aligned}
f_x^m &= \frac{|f_x|}{\frac{y_{OA}}{y_{sum}} \cdot \mu f_z} \Rightarrow 0 \leq f_x^m \leq 1 \\
f_y^m &= \frac{|f_y|}{\frac{x_{OC}}{x_{sum}} \cdot \mu f_z} \Rightarrow 0 \leq f_y^m \leq 1 \\
\tau_x^m &= 2 \times \left(\frac{\left|\tau_x - \frac{(y_{OA} f_z - y_{OB} f_z)}{2}\right|}{y_{OA} f_z + y_{OB} f_z} \right) \Rightarrow 0 \leq \tau_x^m \leq 1 \\
\tau_y^m &= 2 \times \left(\frac{\left|\tau_y - \frac{(x_{OB} f_z - x_{OC} f_z)}{2}\right|}{x_{OB} f_z + x_{OC} f_z} \right) \Rightarrow 0 \leq \tau_y^m \leq 1 \\
\tau_z^m &= 2 \times \left(\frac{\left|\tau_z - \frac{(\tau_{max} + \tau_{min})}{2}\right|}{\tau_{max} - \tau_{min}} \right) \Rightarrow 0 \leq \tau_z^m \leq 1
\end{aligned}
\tag{29}
$$

3. Results

The proposed methodology described in Section 2 was analyzed in simulation and via experimental tests using *Transcend*, a life-size humanoid robot (**Figure** 7) developed by 4Front Robotics Inc. and the University of Calgary.

3.1 Simulation results

Before the proposed stability method was implemented on a real robot, a comprehensive simulation study was performed to evaluate its functionality and analyze and determine its limitations. To conduct the simulations, the commercial-off-the-shelf (COTS) multi-body dynamics simulation software Adams was used from where the stability of the robot of interest was determined. The software Adams enabled us to study the dynamics of the parts of the moving robot and determine how forces and loads are distributed throughout the robot affecting its stability. The results of such simulations in Adams were assumed to be accurate and then used to compare with the results of the proposed method. The consistency between the Adams and results obtained from the proposed methodology were used to determine the validity and accuracy of the proposed method. The performance of the method was assessed according to two complementary measures: (i) accuracy and (ii) reliability.

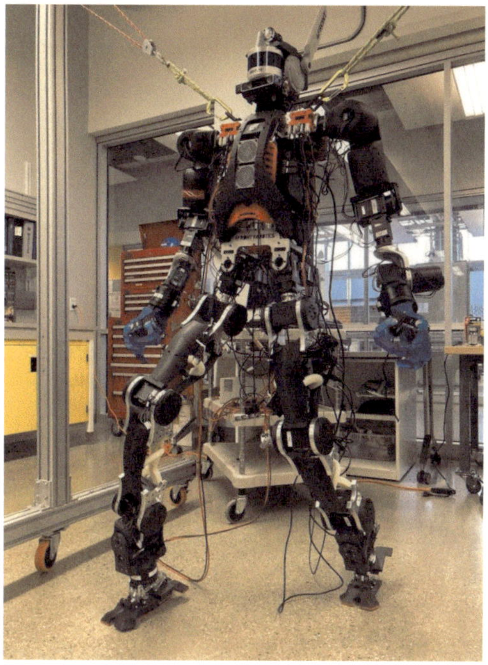

Figure 7.
TRANSCEND: Experimental life-size humanoid robotic platform.

Accuracy is defined as the number of cases when both simulation and theoretical formulation produce identical results under the same robot state (i.e., stable or unstable). Such value is then divided by the total number of cases tested resulting in the *accuracy value*.

Reliability is defined as a measurement of the inconsistency in the results provided by the theoretical and the simulation tests under the same robot state. The purpose of this indicator is to determine how reliable is the proposed method. The inconsistency of the results is considered *reliable* when the mathematical formulation shows that the robot is unstable, but the simulation shows that it is stable. On the other hand, the inconsistency is considered *unreliable* when the mathematical formulation predicts that the robot is stable, while the simulation shows that it is not.

To be able to generate realistic results from the simulation that can be experimentally tested in a reproducible fashion, four terrain characteristics were used: (*i*) coefficient of friction, (*ii*) damping ratio, (*iii*) terrain stiffness, and (*iv*) penetration depth. These characteristics were defined on the contact surface between the robot's foot and the ground. Obviously, such characteristics are dependent on the materials of both surfaces (the ground and the robot's sole/foot), which was known (the robot and the terrain were engineered). The value of each of these four parameters used in the experimental tests was selected based on prior work in the Adams software [19] (see **Table 2**).

During each experimental test, the robot was positioned on a specific surface having four of the properties outlined in **Table 2** (i.e., terrain stiffness friction, and damping). In turn, the robot was positioned in a given (selected) kinematic state after which the corresponding simulation was performed. Each simulation test was conducted three times and each simulation used a different penetration depth coefficient. The reason for such a testing scenario was to investigate the extent to which the

Parameter	Value
Terrain stiffness	$10^9\ N/m$
Coefficient of friction	0.3
Damping	$6.5 \times 10^3\ N.s/m$
Penetration depth	$0, 10^{-4}, 2.1 \times 10^{-3}\ mm$

Table 2.
Contact characteristics used.

proposed methodology can be used to determine the robot's stability when standing on compliant (deformable) surfaces and not just solid (rigid) terrains. Although numerous surface conditions (e.g., slope, friction, etc.) were used during the experimental tests, for simplicity, in the context of this paper only three representative horizontal flat surfaces are used to exemplify the obtained results: *surface 1*, *surface 2*, and *surface 3*.

Surface 1: A solid surface with a penetration depth of zero.

Surface 2: A semi-solid surface with a penetration depth of 10^{-4}, (and)

Surface 3: A compliant surface with a penetration depth of 2.1×10^{-3}.

During each simulation case, the values of the contact forces, f_x^i and f_y^i, at the corresponding contact point C_i were varied from 0 to 100 N between tests. **Figure 8** illustrates the location of the three contact points on the robot's sole used during the simulation and experimental tests. In the experimental tests such contact points were achieved by placing rubber bumpers on the robot's feet at the locations shown in **Figure 8**.

Figure 9 illustrates the robot's stability on different values for f_y^2 on the three different surfaces. The bars denote the binary stability condition obtained from the proposed method (orange) and the simulation (blue bars).

From **Figure 9**, it is observed that the proposed methodology agrees 100% with the full-body dynamic simulation tests except in the case of surface 3, when a force $f_y^2 = 10N$ was applied. Under such conditions, the simulations and proposed mathematical

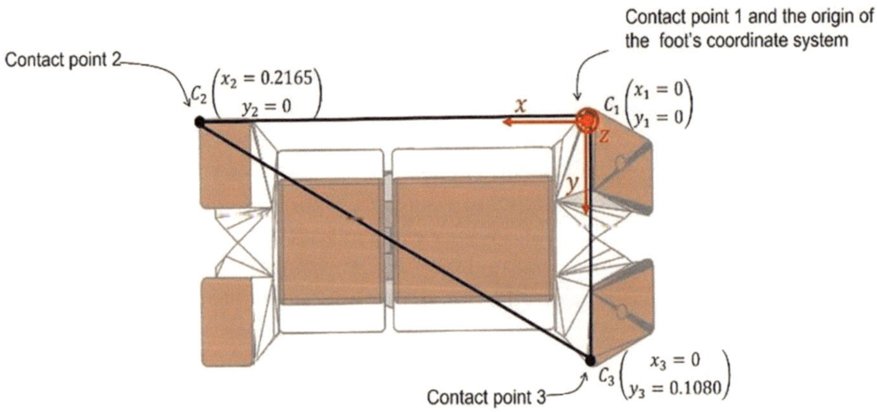

Figure 8.
Location of the pre-defined three contact points ($\mathbf{C_1}$, $\mathbf{C_2}$, and $\mathbf{C_3}$) on the robot's sole.

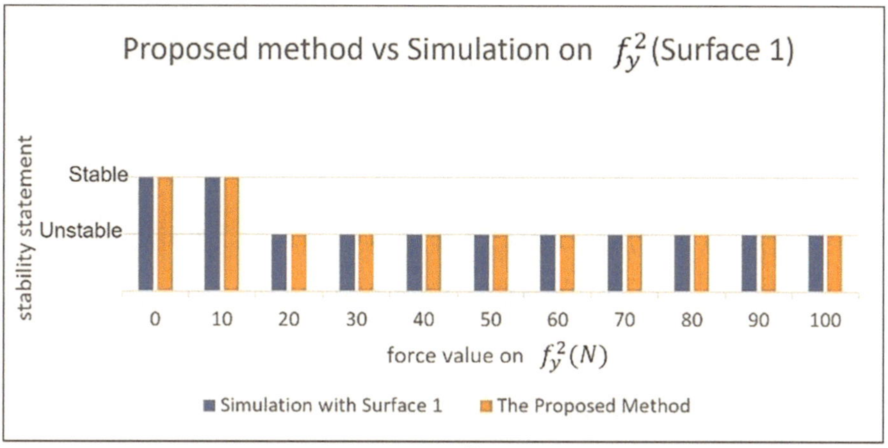

a) Stability of the when $f_y^2(N)$ is varied on Surface 1.

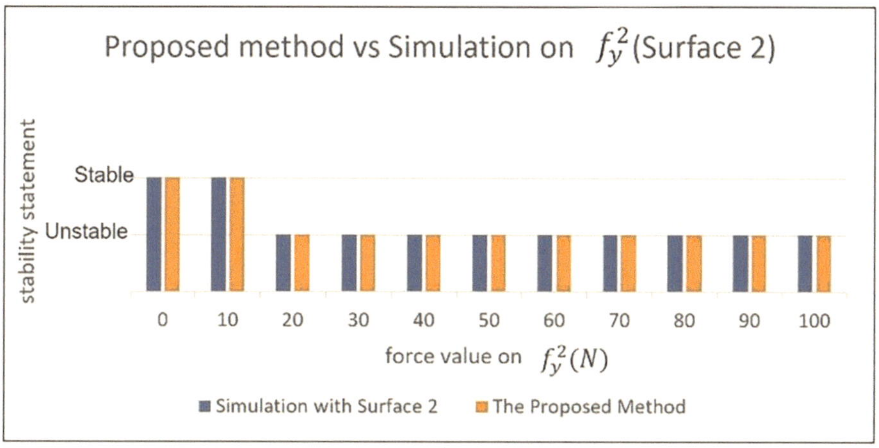

b) Stability of the when $f_y^2(N)$ is varied on Surface 2.

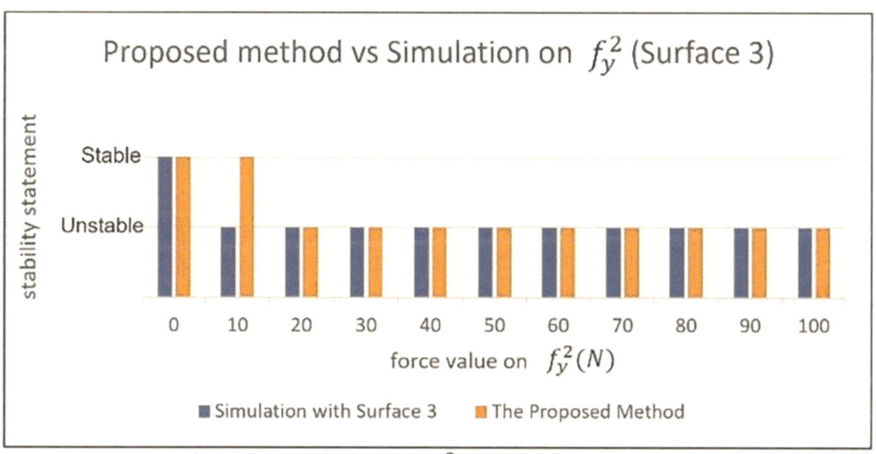

c) Stability of the when $f_y^2(N)$ is varied on Surface 3.

Figure 9.
Stability results during the simulation as the robot stand on a flat horizontal surface when one of the contact forces (f_y^2) is varied.

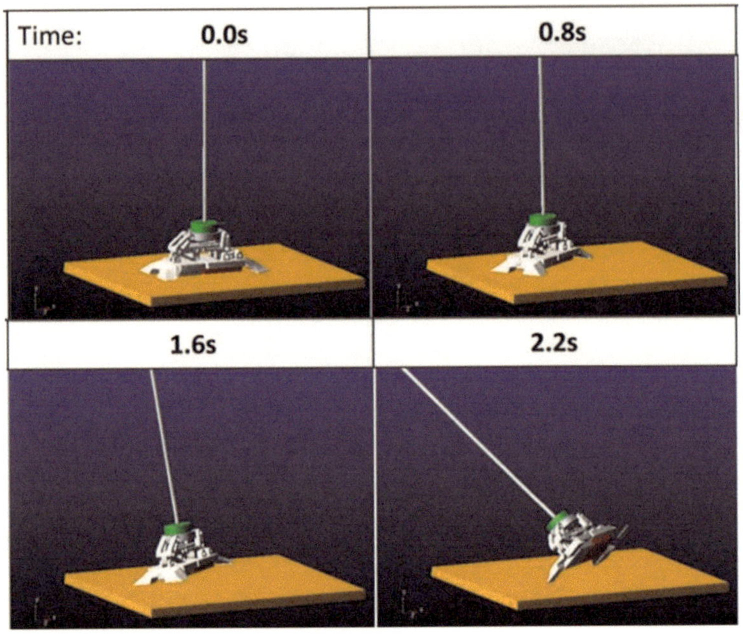

Figure 10.
Visualization of the foot's motion when in an unstable condition.

formulation indicate that the robot is in an unstable and stable state, respectively. This discrepancy is thought to be due to the fact that the simulated full-body dynamics is able to identify that the robot's foot is sliding (due to the low friction that exist at the contact points) while experimentally such aspects were sometimes difficult to easily visualize (**Figure 9**). **Figure 10** shows the corresponding motion of the robot's foot when sliding on the surface and taken at different time steps causing the robot to lose stability and fall.

The same procedure was also performed for the three components of the three contact points, f_x^2, f_x^3, f_y^1, f_y^2, and f_y^3 while varying the corresponding force caused by gravity at each simulation on a specific surface. Such tests produced similar results to the results shown in **Figure 9**. The results of these tests in terms of their accuracy and reliability are presented in **Tables 3–5** for the specific surfaces used (i.e., solid, semi-solid, and compliant).

The obtained results show that the proposed method is successful in estimating the stability of the robot on all surfaces. With almost the same accuracy for the solid and semi-solid surfaces (**Tables 3** and **4**, respectively). Also, the method shows reliable

Surface #1	**Contact point 1**		**Contact point 2**		**Contact point 3**	
	f_x^1	f_y^1	f_x^2	f_y^2	f_x^3	f_y^3
Accuracy (%)	72.7	100	72.7	100	81.8	100
Reliability (Y/N)	Y	Y	Y	Y	Y	Y

Table 3.
Performance of the proposed method compared to the simulation on a flat horizontal surface 1.

Surface #2	Contact point 1		Contact point 2		Contact point 3	
	f_x^1	f_y^1	f_x^2	f_y^2	f_x^3	f_y^3
Accuracy (%)	72.7	100	72.7	100	81.8	100
Reliability (Y/N)	Y	Y	Y	Y	Y	Y

Table 4.
Performance of the proposed method compared to the simulation on a flat horizontal surface 2.

Surface #3	Contact point 1		Contact point 2		Contact point 3	
	f_x^1	f_y^1	f_x^2	f_y^2	f_x^3	f_y^3
Accuracy (%)	100	90.9	90.9	90.9	100	81.8
Reliability (Y/N)	Y	N	Y	N	Y	N

Table 5.
Performance of the proposed method compared to the simulation on a flat horizontal surface 3.

results in all cases tested, which suggests that the proposed method can accurately predict all unstable conditions. The results show that the proposed methodology provides better results on solid surfaces and surfaces having small deformation (**Table 5**).

Figure 11 shows the corresponding radar charts for the test shown in **Figure 9a**. **Figure 11a** shows the robustness status of the robot when all contact force components are equal to zero (except for the vertical forces, f_z, and caused by the robot's weight). The obtained radar chart 10a shows a "highly stable" condition as all five parameters are within the green section of the chart. **Figure 11b**, as an unstable example, shows the case when the force f_y^2 was increased to 10 N. **Figure 11c** shows the case when the humanoid robot is in a "moderate-risk stability" region despite the fact that the force f_y^2 was increased to 15 N when compared to previous tests. Lastly, when the force f_y^2 was increased to 20 N, the robot is seen to lose stability. That is, the generated data polygon crosses the borders of the stability polygon where the robustness concept is not defined (**Figure 11d**).

The proposed approach was also tested on slope surfaces. To this aim, three surfaces having slopes of 10, 20, and 30 degrees were used. The simulation settings used to test the stability on these surfaces were the same as provided in **Table 2** and **Figure 8**.

The results on tilted surfaces having an angle between 10 and 20 degrees produced reasonable results with a stability accuracy value between 72 and 100%. Similar to the simulated tests on flat horizontal solid and semi-solid surfaces, the results are considered to be reliable. However, some of the results obtained for compliant surfaces were unreliable [18]. When increasing the surface slope to 30 degrees, it was observed that under the same settings used for the 20° sloped terrain, both the simulation and the proposed method show that the robot losses stability as expected without the robot receiving help via external forces (e.g., supporting itself from a banister of from any other surface with its hand).

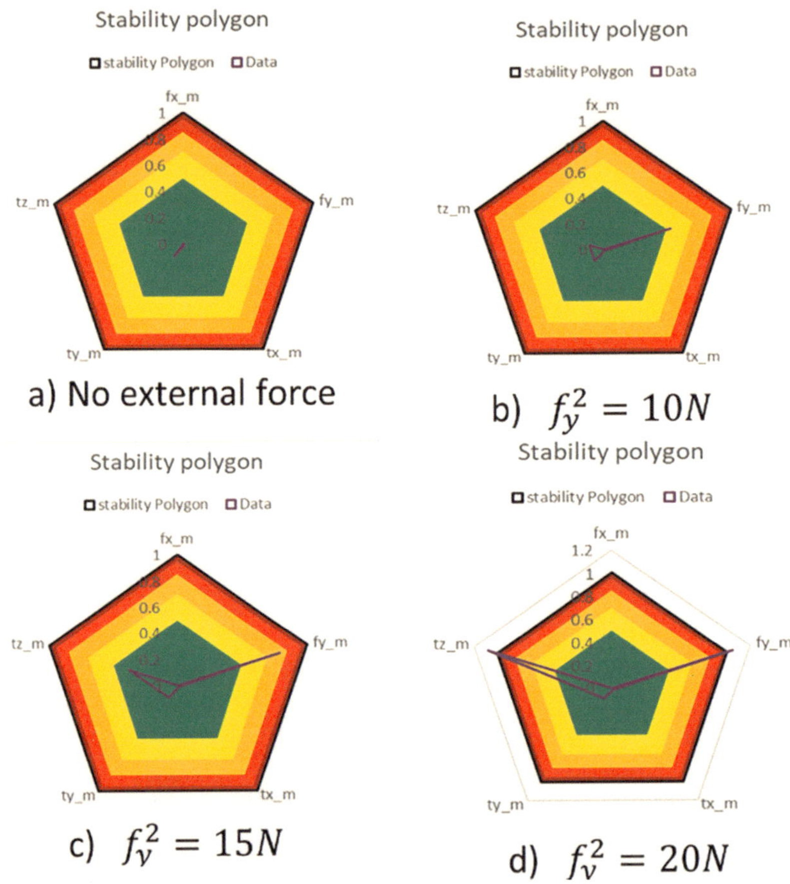

Figure 11.
Robustness check for different force values of f_y^2 on a flat horizontal surface.

3.2 Experimental tests

After analyzing and validating the proposed method in simulation, the method was experimentally tested on real humanoid hardware to examine the proposed method under real-world conditions. To test the proposed stability mechanism the humanoid robot, *transcend* (**Figure** 7) was used. To enable such tests a ROS (robot operating system) package was developed encoding the proposed formulations and configured to read sensor data from the robot's F/T sensors in real time. **Table 6** shows the pseudo-code for the developed ROS package.

To check the stability when standing on slanted surfaces, the robot was positioned on the corresponding surface standing on one of its feet. Then, the ROS package was executed, and the stability results were published continuously. The robot was placed on a flat rigid surface as well as on a bumpy sloped terrain. Once the robot was in place the slope value of the terrain was changed as the robot kept its state (did not attempt to maintain balance).

It was observed that when the stability condition of the robot changes, the proposed method is able to accurately perceive such changes and compute the stability of the robot while publishing the updated stability condition in real time. After numerous experimental tests on various terrain conditions, it is concluded that the proposed

Inputs	Algorithm
• Wrench at the location of the F/T sensor for each timestamp comprising three force values and three torque values: $w_{inst} = \begin{bmatrix} f_x & f_y & f_z \\ \tau_x & \tau_y & \tau_z \end{bmatrix}$ • Friction coefficient value (μ). • The location of the contact points. **Outputs** • The stability condition for each timestamp. **Parameters used in the algorithm:** • f_x,f_y,f_z: The force values in the x-axis, y-axis, and z-axis, respectively, at the origin of the F/T sensor's coordinate system. • τ_x, τ_y, τ_z: The torque values in the x-axis, y-axis, and z-axis, respectively, at the origin of the F/T sensor's coordinate system.	**Step 1:** Read the wrench data published by the F/T sensors (w_{inst}). **Step 2:** Calculate the upper and lower boundaries of the six variables (f_x,f_y,f_z,τ_x, τ_y, and τ_z) based on the given friction coefficient and predefined location of the contact points. **Step 3:** Check whether the values of w_{inst} lie inside the specified boundaries obtained in step 2. • If all values of w_{inst} lie inside their corresponding related boundaries, then the robot is ***stable***. • If one of w_{inst} values do not lie inside their corresponding boundaries, then the system is ***unstable***. **Step 4:** Publish the results of step 3.

Table 6.
The pseudocode of the ROS package was developed to test the robot's stability.

method accurately predicts the robot's stability conditions on various surfaces having different friction coefficients, slopped conditions, and levels of surface characteristics, such as diverse bumpiness conditions.

4. Discussion

With an increasing demand for utilizing robotic tools in many hazardous/labor-intense working environments, such as mining and forestry, the need for obtaining a comprehensive and practical robotics approach using humanoid robots becomes essential. Although remarkable studies have been conducted to enable humanoids to execute tasks in environments engineered for humans, one of the biggest problems prohibiting the full integration of these systems in complex labor-intense working environments is that they are incapable of successfully maneuvering through such complex chaotic spaces where they must maintain their stability while performing operations. The main reason for this incapability is the lack of a universal (yet fast) solution to the problem of humanoid stability on complex and unstructured terrains. Most of the relevant studies conducted on humanoids' stability have focused on humanoids walking on flat horizontal terrains which do not represent many real-world scenarios (e.g., urban search and rescue operations inside collapsed infrastructure). Prior work on multi-legged robot stability including the multi-contact approaches, such as the CWS and the ZMP methods, are limited as they do not provide effective solutions for cases where the robot's feet contact the environment on multiple noncoplanar surfaces.

The solution presented in this chapter allows humanoids to use not only their feet but also any other parts of their body to contact the environment to enhance their stability and hence improve their locomotion. In this regard, the modified version of CWS discussed in this work provides a generic approach that extends the capabilities of the ZMP and the traditional CWS approaches. Although the modified CWS method provides an effective procedure for determining the stability of robots in multi-contact locomotion procedures, there are several improvements that can be conducted to further enhance the approach. Some of these improvements include:

- Develop suitable feedback controllers using stability, such as fuzzy feedback techniques, to control the robot's forces as well as torques in such a way that robot always stays in a stable condition. For this, improved position controllers combined with force/torque-based controllers should be formulated.

- With the presented stability formulation planning, the path of a humanoid has the capability to be enhanced as the formulation could provide useful information to determine the robot's stability based on future movements as opposed to simply determine if the robot is stable but rather determining the robot's stability with respect to the various motions that the robot could execute.

Although the presented stability method provides a rigorous theoretical framework, the method would benefit from a detailed experimental comparison with respect to previously formulated stability determination approaches to precisely quantify its superiority in the real-world scenario, hence improving human-robot collaboration. Human-robot collaborative systems (HRCS) have emerged as a solution that aims to satisfy the current demand for highly personalized products, enhance quality, and reduce costs by combining the skills of humans and robots within a shared workspace. However, the safety of human operators, and people in general, is of major concern when located in collaborative workstations and/or common areas. In this regard, it is envisioned that the methodology presented in this manuscript can significantly improve safety assurance practices in diverse sectors, including current applications of the manufacturing industry [20]. In this sense, the integration of human-robot interactions and safety standards has been analyzed [20], where a favoring inclination to techno-centered practices has been noticed, where policies designed for rigid automation are being used for HRCS alike. This, however, brings additional risks to the well-being of human operators.

Industry 4.0, as the current transformation of manufacturing systems, introduces a paradigm shift supported by new robotic technologies that together enable the decentralization of production processes via HRCS [21, 22]. For instance, collaborative robots (or cobots) have been developed with intuitive interfaces, sensors, and software to aid human operators in performing demanding physical tasks (i.e., handling dangerous materials or executing repetitive operations) [23]. Furthermore, since HRCS aims to achieve higher flexibility in the production of diverse-sized volumes, HRCS is being adopted in the manufacturing and other industries as a solution that mixes, within a shared workspace, the accuracy and repeatability skills of collaborative robots with the dexterity and cognitive capabilities of human operators. This semi-automated schema, which integrates modern robotics and legacy manual work, is a promising instrument for mass customization goals [24] in which the methodologies presented in this manuscript will find useful applications.

5. Conclusions

The proposed theoretical formulation based on the CWS stability approach was found to be applicable to any continuous (e.g., floor or loose dirt) or discrete (e.g., sharp rocky surface) contact surface that can be represented by an arbitrary set of three contact points. The proposed approach provides the ability for a fast and efficient stability computation for multi-legged robots operating in a broad spectrum of applications. Hence, it provides opportunities to deploy humanoids and other types of robots in various real-world environments as complex as the environments encountered in hazardous urban search and rescue operations. The developed approach enables fast and efficient stability computation that can be utilized in a wide range of applications, such as hazardous spaces, including moving inside collapsed buildings. The introduced formulations equip roboticists with a superior stability tool that will enable them to better control robots walking in complex potentially dynamic environments. The results from simulations and experimental tests show that the proposed method is a powerful tool capable of computing stability in a very fast way with reasonable accuracy and reliability. As a result, the proposed approach provides relevant information that can be used to control the body motions of multi-legged robots so that their actions do not compromise their stability.

Author details

Parastoo Dastangoo and Alejandro Ramirez-Serrano*
Department of Mechanical and Manufacturing Engineering, University of Calgary, Calgary, Canada

*Address all correspondence to: aramirez@ucalgary.ca

References

[1] Vukobratović M, Borovac B. Zero-moment point—Thirty five years of its life. International Journal of Humanoid Robotics. 2004;**1**(01):157-173

[2] Takao S. Yokokohji Y. Yoshikawa T. FSW (feasible solution of wrench) for multi-legged robots. In: 2003 IEEE International Conference on Robotics and Automation. IEEE. 2003;**3**: 3815-3820

[3] Dai H. Robust multi-contact dynamical motion planning using contact wrench set [A thesis]. Massachusetts Institute of Technology. 2016

[4] Wieber P-B. On the stability of walking systems. In: Proceedings of the International Workshop on Humanoid and Human Friendly Robotics. 2002

[5] Hirukawa H. Humanoid robotics projects in Japan. In: Proceedings of the Understanding Humanoid Robots Workshop at IEEE ICRA 2006. 2006

[6] Dai H, Valenzuela A, Tedrake R. Whole-body motion planning with centroidal dynamics and full kinematics. In: 2014 IEEE-RAS International Conference on Humanoid Robots. IEEE. 2014. pp. 295-302

[7] Koyanagi Ki, et al. A pattern generator of humanoid robots walking on a rough terrain using a handrail. In: 2008 IEEE/RSJ International Conference on Intelligent Robots and Systems. IEEE. 2008. pp. 2617-2622

[8] Caron S, Pham Q-C, Nakamura Y. Leveraging cone double description for multi-contact stability of humanoids with applications to statics and dynamics. Robotics: Science and Systems. 2015;**11**.1-9

[9] Escande A, Kheddar A, Miossec S. Planning contact points for humanoid robots. Robotics and Autonomous Systems. 2013;**61**(5):428-442

[10] Herzog A, Rotella N, Schaal S, Righetti L. Trajectory generation for multi-contact momentum control. In: 2015 IEEE-RAS 15th International Conference on Humanoid Robots (Humanoids). IEEE. 2015. pp. 874-880

[11] Zheng Y, Lin MC, Manocha D, Adiwahono AH, Chew C-M. A walking pattern generator for biped robots on uneven terrains. In: 2010 IEEE/RSJ International Conference on Intelligent Robots and Systems. IEEE. 2010. pp. 4483-4488

[12] Wang Z, Harada K, Wan W. Multi-contact Stability of Humanoids using ZMP and CWC. In: 2019 IEEE-RAS 19th International Conference on Humanoid Robots (Humanoids). IEEE. 2019. pp. 126-131

[13] Caron S, Pham Q-C, Nakamura Y. Stability of surface contacts for humanoid robots: Closed-form formulae of the contact wrench cone for rectangular support areas. In: 2015 IEEE International Conference on Robotics and Automation (ICRA). IEEE. 2015. pp. 5107-5112

[14] Wiedebach G, Bertrand S, Wu T, Fiorio L, McCrory S, Griffin R, et al. Walking on partial footholds including line contacts with the humanoid robot atlas. In: 2016 IEEE-RAS 16th International Conference on Humanoid Robots (Humanoids). IEEE. 2016. pp. 1312-1319

[15] Kagami S, Kitagawa T, Nishiwaki K, Sugihara T, Inaba M, Inoue H. A fast dynamically equilibrated walking

trajectory generation method of humanoid robot. Autonomous Robots. 2002;**12**(1):71-82

[16] Caron S. Computational foundation for planner-in-the-loop multi-contact whole-body control of humanoid robots [A thesis]. University of Tokyo. 2016

[17] Trinkle JC, Pang JS, Sudarsky S, Lo G. On dynamic multi-rigid-body contact problems with Coulomb friction. ZAMM-Journal of Applied Mathematics and Mechanics/Zeitschrift für Angewandte Mathematik und Mechanik. 1997;**77**(4):267-279

[18] Dastangoo P. Multi-contact stability of multi-legged robots operating in unstructured spaces [MS thesis]. Schulich School of Engineering. 2021

[19] Daniel WW. MSC.Software. Contact modeling presentation, sheet 10. Available from: https://www.insumma.nl/wp-content/uploads/SayField_Verheul_ADAMS_Contacts.pdf

[20] Segura Parra P, Lobato-Calleros O, Ramirez-Serrano A, Soria I, Human-robot collaboration systems: Structural components for current manufacturing applications, Journal of Advances in Industrial Manufacturing Engineering. Nov 2021;**3**:114-125

[21] Costa Mateus JE, Aghezzaf E-H, Claeys D, Limère V, Cottyn J. Method for transition from manual assembly to Human-Robot collaborative assembly. IFAC-PapersOnLine. 2018;**51**(11): 405-410

[22] Mateus JEC, Claeys D, Limère V, Cottyn J, Aghezzaf E-H. Ergonomic and performance factors for Human-robot collaborative workplace design and evaluation. IFAC-PapersOnLine. 2019; **52**(13):2550-2555

[23] Parra PS, Calleros OL, Ramirez-Serrano A. Human-robot collaboration systems: Components and applications. In: International Conference of Control, Dynamic Systems, and Robotics, 2020. 2020. pp. 1-9

[24] Ding Y, Xu W, Liu Z, Zhou Z, Pham DT. Robotic task oriented knowledge graph for human-robot collaboration in disassembly. Procedia CIRP. 2019;**83**:105-110

Chapter 6

Perspective Chapter: Dielectric Elastomer Sensor Capable of Measuring Large Deformation and Pressure

Seiki Chiba and Mikio Waki

Abstract

Most of the conventional sensors used for measuring deformation, pressure, etc., use metal, ceramics, piezo, or the like. Many of them are very rigid, and when the object is deformed or when the pressure on the object changes currently, it is necessary to arrange a large number of sensors with different conditions side by side. However, it is still difficult to measure all changes over time. With the newly developed dielectric elastomer sensor, even a very thin (0.1–0.2 mm) elastomer thickness could be deformed in difficult environments (e.g., places with large temperature changes or large vibrations), and it would be possible to measure any pressure changes due to its deformation. By applying this sensor, it can be used as a position sensor (including a three-dimensional sensor) or an acceleration sensor, so that it could be applied to the control of the arms and legs of a robot, smart shoes, and the like.

Keywords: dielectric elastomer, sensor, pressure, electrodes, CNT, carbon black, carbon grease, load cells, resin block, flange, sheet, SS curve

1. Introduction

Materials consisting of new metal materials, high-performance polymer materials, fine ceramics, and composite materials are expected to be used in cutting-edge technologies that support various industries and economies in the twenty-first century, along with electronics and biotechnology [1].

Piezoelectric composite materials [2], conductive polymer materials [3], bimetals [4], etc., are known as composite materials that can be applied as sensor materials. The former two are excellent materials that combine the advantages of both organic materials with inorganic materials in a matrix. The latter combines dissimilar metals to enable temperature sensors.

Engineering sensors are extremely simple devices at the present stage. As seen in partially integrated pressure sensors and magnetic sensors, they are becoming more intelligent, but they have not yet reached the level of intelligent sensors. As an example of the development of an integrated pressure sensor, T. Sarutani et al.

prototyped an integrated pressure sensor using a shear gauge as a piezo resistance gauge [2]. The temperature of the pressure sensor is corrected on the same chip. As a feature of this sensor, the pressure that can be measured is only the pressure at the time when the pressure is applied, and accurate measurement cannot be performed unless the temperature is corrected.

As a sensor for measuring elongation, a strain sensor can be mentioned. The type using a piezo is well known [5]. However, the measurement of elongation is quite limited. Stretch sensors that support greater elongation using elastomers have recently emerged. Electrodes were attached to the top and bottom of the elastomer, and the elements were stacked in several stages for use [6]. It is an application of the so-called dielectric elastomer (DE). However, since a plurality of these elements are laminated, it is considered that the flexibility is not so high, so it is considered that the element is not sufficiently deformed unless the force required for deformation is large.

In this experiment, carbon grease, carbon black, and single-wall carbon nanotubes (SWCNTs) were applied to thin elastomer films of silicon, acrylic, and hydrogenated nitrile rubber (HNBR) as electrodes, and each film was used as a pressure sensor. Then, how each film behaved as a pressure sensor was observed. We also verified the performance as a stretch sensor by using the same combination of film and electrodes as above. In this way, we discussed whether a film with the same structure could be used as an intelligent sensor capable of two different types of sensing.

2. Background of dielectric elastomers (DEs)

DE was first created in 1990 by S. Chiba, R. Pelrine et al. of the Stanford Research Institute in the United States. After that, various researchers started their own development [7–28]. Currently, that development is spreading to DE actuators (DEA) and DE power generators (DEGs) that generate electricity by reversing the drive [28]. The concept of the DE sensor was introduced in the latter half of the 1990s, and research and development are currently underway in this field as well [8–25].

The structure and principle of the DE sensor (see **Figure 1**) are the same as those of the DEA and DEG, and the driving principle is that the output of DEA and its elongation are in inverse proportion to each other. Also, the change in capacitance of the DE sensor and DEA and its elongation are in direct proportional relationship [9] (see **Figure 2**).

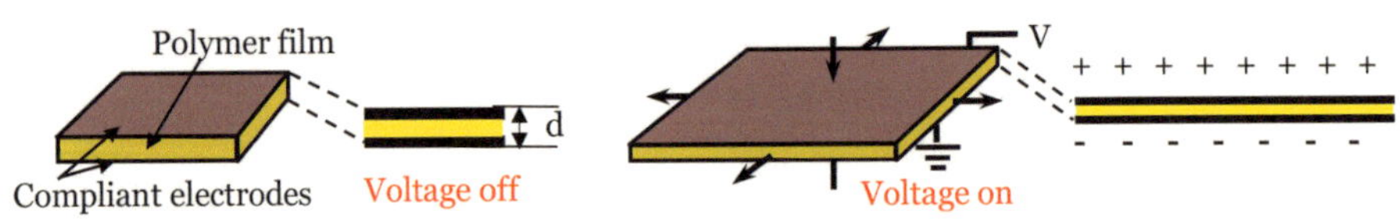

Figure 1.
Structure and principle of DE sensor.

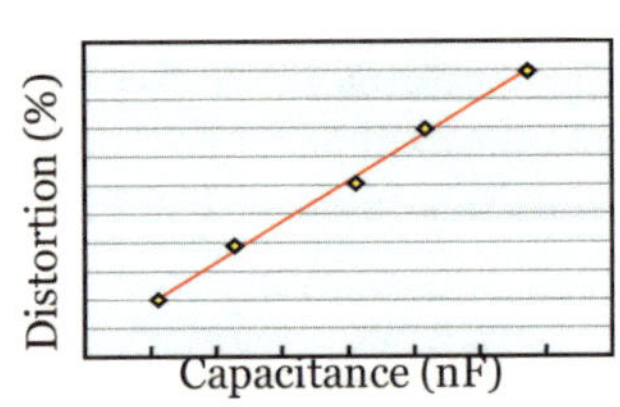

Figure 2.
Relationship between the elongation of the DE sensor and capacitance.

The relationship between DEA and capacitance can be expressed by the following equation;

$$C = \varepsilon \frac{S}{d} \tag{1}$$

Here, C is capacity (F), ε is the dielectric constant of polymer film (F/m), S is area of Electrode (m^2), and d is distance between electrodes (m).

So far, a wide range of research and development have been carried out for various applications. This research also includes the examination of material types and DE sensor shapes [10–24]. First, as given an example of the material, H. Sun et al. claimed that SiR-Fe containing 8% $FeCl_3$ could be effectively used as a medical DE sensor [10]. Perhaps this polymer is hard and unsuitable for precision sensors. B. O'brien et al. made a DE sensor by using two layers of thick silicone elastomer with a size of 20 mm x 70 mm and applying carbon powder to the electrodes [11]. M. Han et al. proposed a pressure sensor in which the medical wearable sensor consisted of a carbon nanotube-polydimethylsiloxane (CNT-PDMS) composite electrode and a porous polymer dielectric layer [12]. The concept is interesting, but this wearable sensor is unlikely to work well because the electrodes were not flexible enough and polymers used were hard.

H. Liebscher et al. tried to increase the dielectric constant of polyurethane by adding barium titanate to the polyurethane elastomer [13]. J. Bae et al. tried to control the robot by feeding back the information [14]. C. Briggs et al. tried to develop a tactile sensor for a robot's hand by attaching a dome-shaped DE sensor or DE sensor mounted on a cylindrical pad to a robot gripper [15]. L. Agostini et al. also conducted preliminary experimental studies seeking the most influential elements in order to optimize the performance of hemispherical anti-collision sensors [16].

As an example of circuit studies, K. Jung et al. used modulation technology to realize DEA operation and its detection and used a system that mixed a low-frequency signal for operation and a high-frequency signal with small amplitude for detection [17]. H. Bose et al. tried to create a thicker mat-shaped pressure sensor because the sensor signal was too small to detect in the sheet-shaped DE [18]. Finally, as an application example, J. Bae et al. also explained applications using PVDF-based materials, silicones, and acrylic materials [14]. M. Rosenthal et al. discussed diagnostic tools for industrial equipment and applicable sensors for system monitoring [19].

R. Walker et al. conducted a unique study of attaching a DE sensor to a wetsuit for divers as an underwater application for DE. [20]. C. Larson et al. also arranged a large number of DE sensors on participants' bodies and tried to apply them to a virtual reality game that would enable them to play the game by moving their bodies [21]. As another system research example, a DEA system could be developed that combines a DE sensor and an actuator to assist the movement of a patient's fingers, hands, feet, etc., and to accurately evaluate their rehabilitation progress [22]. R Venkatraman et al. tried to measure blood pressure using a DEA cuff device [23].

In this way, DEA can be used simultaneously an actuator and a pressure sensor and/or a position sensor [24]. In the near future, it is expected that the arms and legs of intelligent robots, nursing care equipment, and the like will become possible. In addition, a DEA/sensor system that assists the movement of the patient's fingers, hands, feet, etc., and accurately evaluates the rehabilitation situation will be possible [25]. In developing such devices and devices, it is necessary to improve the performance of DE sensors. The major factors will be to improve the conductivity of the electrode and the flexibility of the electrode [26, 27].

3. Experimental procedure

In this experiment, carbon grease, carbon black, and SWCNT (ZEONANO®-SG101, Zeon Corp., Tokyo, Japan) were used as electrodes for four types of elastomers to create DE pressure sensors, and the performance of each was compared. If necessary, soft resin blocks were laid under the elastomer films to further increase the sensitivity of the DE sensor. Similarly, for the stretch sensor, carbon grease, carbon black, and SWCNT electrodes were used for each, and the performance was compared.

3.1 Test materials

In this study, we used a US-made acrylic material (3 M/4905), a US-made acrylic distortion-corrected film, and a silicon material (ELASTOSIL FILM 2030250). In addition, hydrogenated acrylonitrile butadiene rubber (HNBR) film synthesized by Zeon Corp. was used. The thickness of the acrylic material and silicon of the test piece was 0.5 mm, and the thickness of HNBR was 0.2 mm. HNBR is a material that absorbs the vibrations of automobile engines and is too hard as it is, so we optimized the amount of cross-linking agent added and cut a part of the double bond to improve the elongation [26]. Using the above films and resin blocks, a tensile test and dynamic viscoelasticity test were also performed to correlate sensor performance with material differences.

3.1.1 Tensile test

A tensile test of the dumbbell-shaped test piece as shown in **Figure 3** was performed using the Orientech tabletop material tester STA-1150. The displacement velocities were set to 100 mm/min, 200 mm/min, 300 mm/min, and 500 mm/min in order to evaluate the effect of the displacement velocities on the mechanical properties.

3.1.2 Dynamic viscoelasticity test

The dynamic viscoelasticity tester used was an MCR302 rheometer manufactured by Anton Paar, and the viscoelastic behavior of acrylic DE and silicone DE in frequency dependence was investigated. The measurement conditions were a frequency of 0.1–20 Hz, a shear strain of 1%, and a room temperature of 20° C, and the size of the test piece was 25 mm in diameter.

3.1.3 Elastomer hardness measurements

In this study, the hardness of the acrylic film (3 M/4905), the strain-corrected film of the acrylic film, silicon (ELASTOSIL FILM 2030250), HNBR film, and resin block

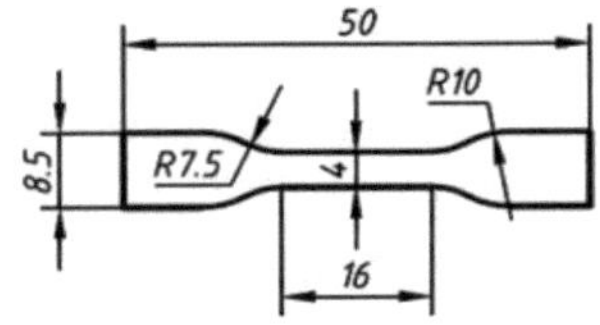

Figure 3.
A tensile test of the dumbbell-shaped test piece.

materials were measured using Askar C method. The details of the resin blocks are shown in Section 3.3.

Askar C is a durometer C (spring type hardness tester) specified in SRIS0101 (Japan Rubber Association standard) and is a measuring instrument for measuring hardness. If Asker C is described as 20 in the physical properties table, it means that the value measured by the Asker C hardness tester is 20, and contrary to the needle insertion degree and consistency test, the larger the number, the harder the material.

3.2 How to make a donut-shaped sheet DE sensor

Each elastomer was cut into a circle and molded to an outer diameter of approximately 8 cm using 100% prestrain. A carbon grease electrode, a carbon black electrode, and an SWCNT electrode were attached to the upper and lower parts of these films to form a donut-shaped sheet DE sensor (see **Figure 4**). Carbon grease was applied as electrodes with a small brush. As the carbon black electrode, a carbon black spray (carbon black dissolved in a solvent, mixed with a small amount of binder, and packed in a spray can) was used. In the case of SWCNT electrodes, the electrodes were attached using SWCNT spray (ZEONANO®-SG101/SWCNT dissolved in a solvent, mixed with a small amount of binder, and packed in a spray can) [25]. The thickness of the carbon grease electrode was 100 μm, and the thickness was confirmed using a double-scan high-precision laser measuring instrument (LT-9500 & LT-9010 M) manufactured by Keyence. The thickness of carbon black was 70 μm and that of SWCNT was 50 μm, and the thickness was similarly confirmed using a double-scan high-precision laser measuring instrument.

3.3 Pressure measurement/evaluation methods using donut-shaped sheet DE sensors

As an evaluation method for the various DE pressure sensors described above, pressure was applied using the evaluation system shown in **Figure 5** with a resin block as a cushioning material (Blue Forest Trading LLC: KG-01) laid under the DE sensor. It was measured using a small precision vise attached to the Z-axis precision stage, and a load is applied from the top of the donut-shaped DE sensor to deform it. The experimental donut-shaped DE sensor is installed at the top of the load cell. The load was measured using a load cell capable of measuring up to 20 kg. The analog signal output from the load cell was taken into the CPU via a dedicated AD conversion IC (HX711). After that, the calibration process was performed by the CPU, and then the measured value was displayed on the LDC. The experimental equipment was calibrated using weights of 10 g, 500 g, 1 kg, 5 kg, and 15 kg, and it was confirmed that the error at each measurement point was within ±1%. The electrostatic capacity of the

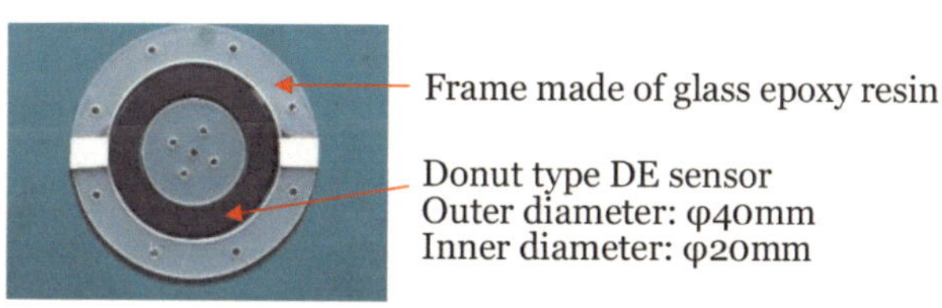

Figure 4.
An example of a donut-shaped sheet DE sensor (using SWCNT electrode).

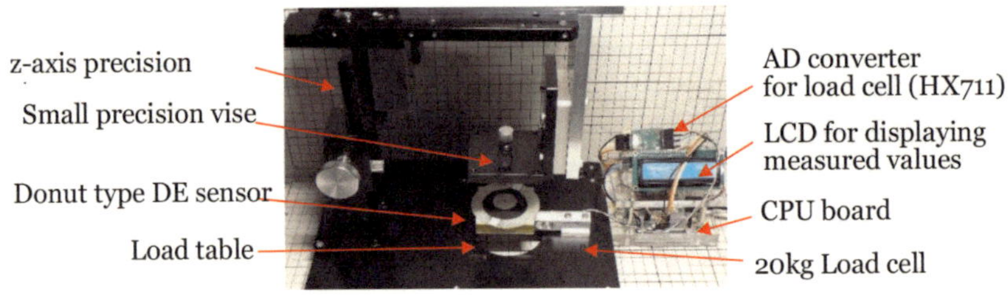

Figure 5.
Overview of electrostatic capacity measuring device for donut-shaped DE sensor.

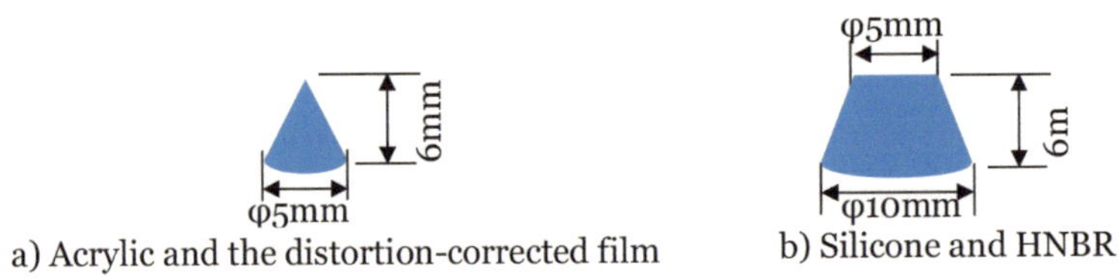

Figure 6.
Donut-shaped DE sensor installed so that the central part rises about 5 mm.

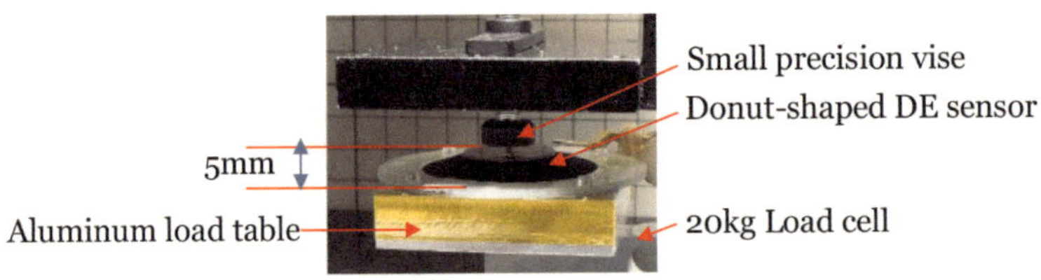

Figure 7.
Shapes of resin blocks used.

donut-shaped DE sensor was measured using LCR METER (ZM2372) manufactured by NF corporation.

As described above, in the donut-shaped DE sensor, a conical resin block made of an acrylic gel sheet was installed in the central part, and the central part was installed so as to be raised by about 5 mm. **Figure 6** shows a donut-shaped DE sensor with a resin block installed in order to rise the central part by about 5 mm. The shape of the resin block is shown in **Figure 7**. The height of the resin block is 6 mm, but this block is quite soft, and when a donut-shaped DE sensor is placed on it, it sinks 1 mm and becomes 5 mm. There are two reasons for inserting this resin block:

- To deform the elastomer more greatly, and

- When using as a carbon grease as an electrode, deformation or leakage will occur if force is applied to the carbon grease layer, so the resin block as a cushioning materials were adapted. Naturally, it had need to be used to disperse the pressure.

For the acrylic and its strain-removed film, the block shown in **Figure 6a** was used, but since silicon and HNBR are hard, it was necessary to enlarge the block as shown in **Figure 6b**. The reason is that with a hard film, the block sinks more than necessary, making it difficult to measure the electrostatic capacity. So, it is necessary to increase the area that receives pressure (see Eq. (1) above).

The response speed of the DE sensor was verified by measuring the time from the time when the load was applied to the DE sensor until the actual capacitance of the DE sensor changed. The presence or absence of a load on the DE sensor was determined by monitoring the amount of deformation of the DE sensor using an ultrahigh-precision laser displacement meter (controller: LC-2400, laser sensor: LC-2440) manufactured by Keyence Co., Ltd.

3.4 DE sheet-type sensors without cushioning materials

In addition to the above experiments, experiments without cushioning material in acrylic sheet sensors, HNBR, and silicon sheet sensors using SWCNT electrodes were executed. The loads were increased from 10 kg to 120 kg at 10 kg intervals, and the capacitance at each load was measured. The electrodes of sensors used were the SWCNT spray described above, and the shape was a circle with a diameter of 20 mm. **Figure 8** shows a sheet DE sensor as the experimental prototype.

Figure 9 shows an outline of the experimental equipment. By sandwiching each prototype sensor in a precision vise and tightening the precision vise, a load is applied to the experimental sensor to deform it. The experimental sensor was attached to a precision vise by sandwiching it between acrylic blocks for insulation. For the load measurement, a load cell capable of measuring up to 120 kg was used. The analog signal from the load cell was taken into the CPU via a dedicated AD conversion IC (HX711). After that, after performing the calibration process with the CPU, the measured value was displayed on the LDC. The measurement accuracy was confirmed at three points of 10 kg, 30 kg, and 50 kg using a weight whose mass was measured in advance, and it was confirmed that the error was 5% or less at each point. The electrostatic capacity that changes due to the deformation of the experimental sensor was measured using LCR METER (ZM2372) manufactured by NF corporation.

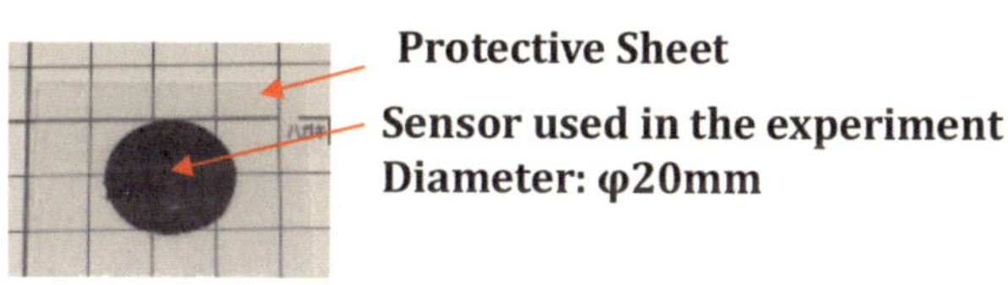

Figure 8.
An experimental sensor with a diameter of 20 mm.

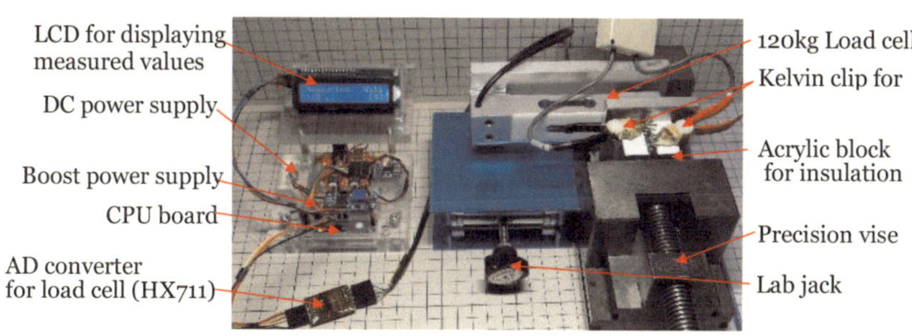

Figure 9.
Outline of the experimental device for sheet-type sensors.

Next, using a DE sheet sensor with a SWCNT electrode attached to a US-made acrylic distortion-corrected film, the pressure was applied up to 120 kgf by sandwiching it between flanges, and its behavior was observed. The meaning of this experiment is that when bolting the flange, it can be confirmed by the pressure sensor whether or not the bolts are tightened evenly. **Figure 10** shows the installation of DE sensors on the flange. **Figure 11** shows sensor bases with four load cells. **Figure 12** shows a pressure change measurement system by tightening bolts using DE sensors.

Acrylic DE sensors (strain removal type) using SWCNT electrodes were installed at four locations on the flange. Capacitance was measured while tightening only the lower right bolt (No. 1) (see **Figure 12**) by 1/2 or 1 turn compared with the others. A calibration was performed in advance by comparing with the value of the DE sensor using a load cell with a maximum measured value of 120 kg (see **Figure 13**). The flange was fixed with four M6 (φ6 mm) bolts attached at 90° intervals. The DE seat sensor was attached to the top of the aluminum plate so that it was located right next to the flange fixing bolt. A glass epoxy plate having a thickness of 0.3 mm is installed on the surface of the aluminum plate for the purpose of electrically insulating the DE sensor and the aluminum block (see **Figure 12**).

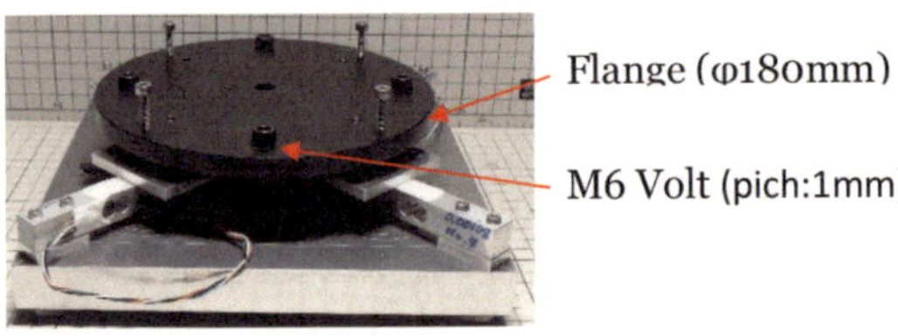

Figure 10.
Installation of DE sensors on the flange and flange fixed with M6 bolts.

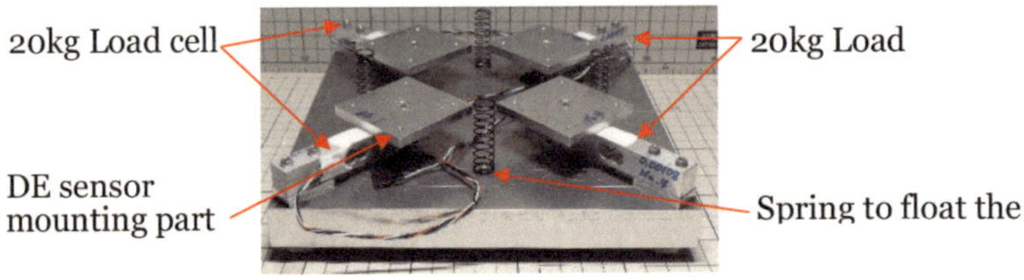

Figure 11.
Sensor bases with four load cells.

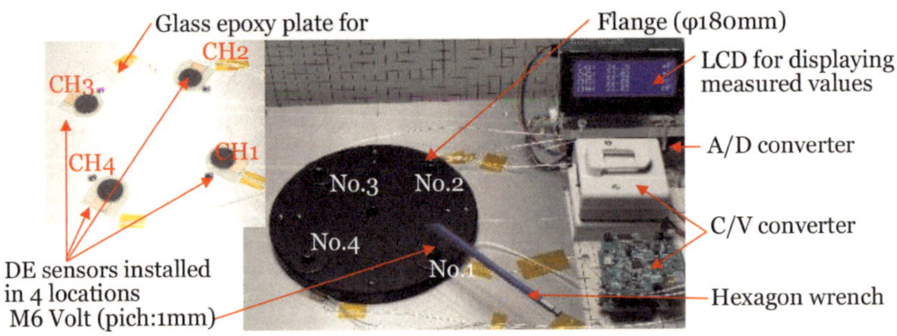

Figure 12.
Pressure change measurement system by tightening bolts using DE sensor.

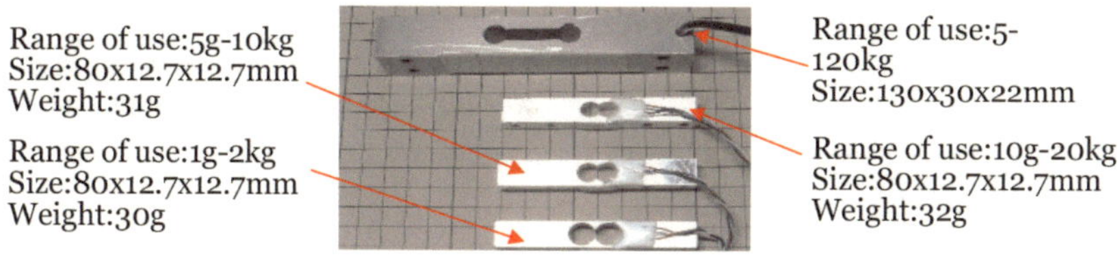

Figure 13.
Examples of weighing scales using various load cells capable of measuring up to 120 kg.

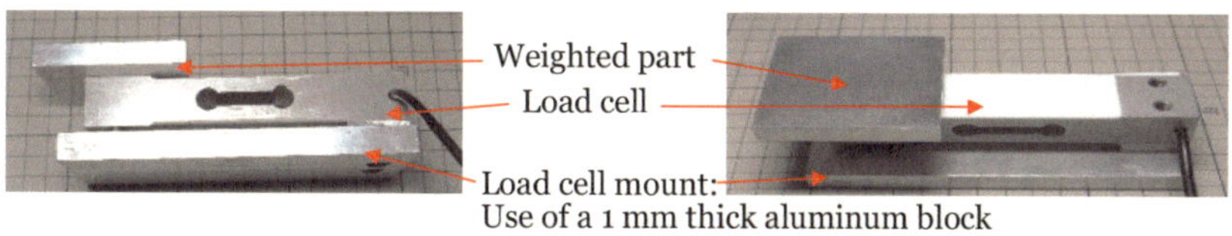

Figure 14.
Load cell and load cell mount.

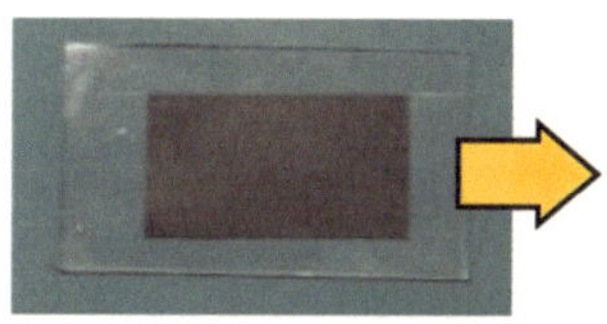

Figure 15.
Sample photograph of a stretch sensor (using carbon grease electrode): Measure the change in conductivity while increasing the stretch by 0.5 cm in the right direction of this sample.

As a circuit to take the data from the DE sensor into the PC, four sets of circuits in which the DE sensor was connected to the CV conversion and the voltage adjustment circuit were prepared. The output from each amplifier was AD-converted and taken into the PC.

Figure 13 shows examples of weighing scales using various load cells capable of measuring up to 120 kg. The load cell alone cannot be used, and a mounting base, weighted part, etc., are required. For example, the weight of a load cell that can weigh up to 120 kg is 200 g, but when the mounting base and weighted parts are added, it becomes about 600 g (see **Figure 14**). In addition to the above, the load cell for measuring several kg has a size of 80 x 12.7 x 12.7 mm and weighs about 30 g. Moreover, in order to use the load cell, a structure that supports the fulcrum with strength that does not deform even with the maximum load is required, which makes it even larger and heavier.

3.5 Manufacturing method of stretch sensor

The elastomer, which had the best performance with the pressure sensor, was cut into a rectangle, molded into a width of 3 cm and a length of 5 cm using 100% prestrain, and an electrode with a width of 2 cm and a length of 3 cm was applied to the center (see **Figure 15**). In order to confirm the difference in performance due to the difference in electrodes, general carbon grease, carbon black, and SWCNT were used as electrodes. For the method of making carbon grease, carbon black, and SWCNT electrodes and the methods of checking those thicknesses, refer to 3.2: How to make a donut-shaped sheet DE sensor.

3.6 Performance comparison by stretch sensor elongation

The change in conductivity was measured while increasing the elongation by 0.5 cm for the sensor using carbon grease, the sensor using carbon black, and the sensor using SWCNT (see **Figure 15**).

4. Results

The above various experiments were performed, and the following results were obtained.

4.1 Pressure measurement result by donut-type sheet DE sensor

The pressure measurement results of the donut-shaped sheet DE sensors prepared using three types of electrodes and four types of elastomer materials are shown in **Figure 16** (carbon grease electrodes), **Figure 17** (carbon black electrodes), **Figure 18** (SWCNT electrodes), and those are summarized in **Table 1**. As shown in **Table 1a**, the silicon material (ELASTOSIL FILM 2030250) has a larger maximum measurable pressure than others. The acrylics have a smaller minimum measurable pressure value than silicon and HNBR. The acrylic with removed distortion and the acrylic with remaining distortion, the measuring minimum pressure value of the removing distortion was about 2 g smaller.

The difference in electrodes did not significantly affect the measurement range. Furthermore, as in the above, there was not much difference in the detection speed due to the difference in the electrodes. Silicon is relatively faster than other membranes (see **Table 1d**). The DE sensor using the SWCNT electrode gave slightly better results than the others. The DE sensor using the SWCNT electrode gave slightly better results than the others. The circuit used for the experiment was fixed to the circuit shown in **Figure 5** for the experiment.

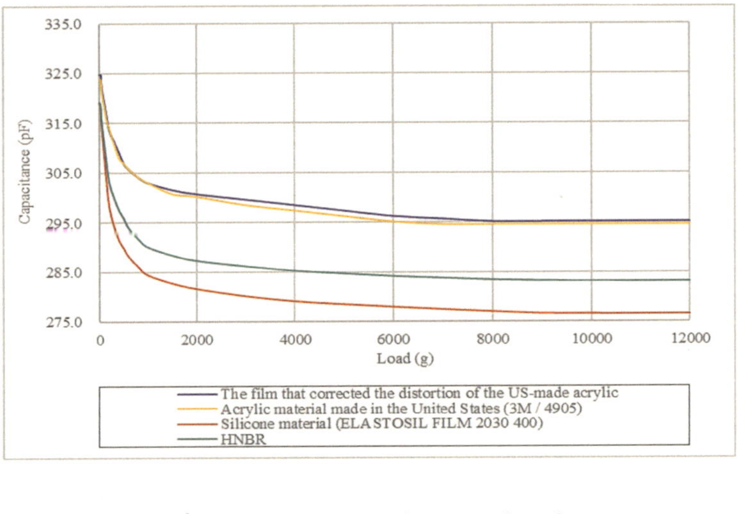

Figure 16
Changes in electrostatic capacities due to load when using carbon grease electrodes

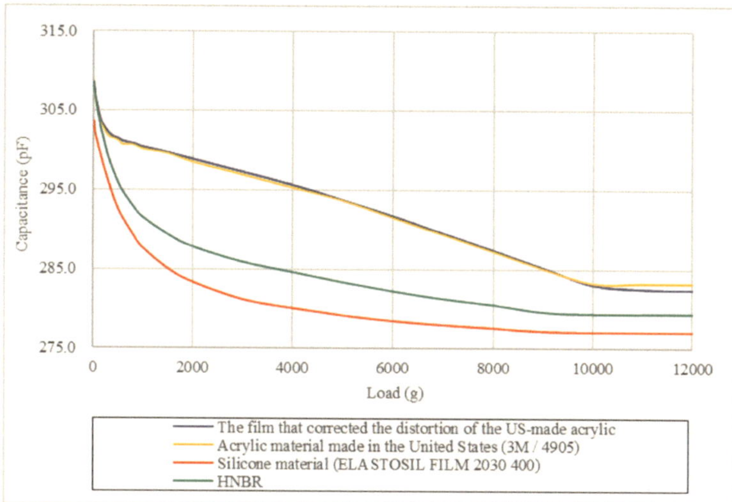

Figure 17.
Changes in electrostatic capacities due to load when using carbon black electrodes.

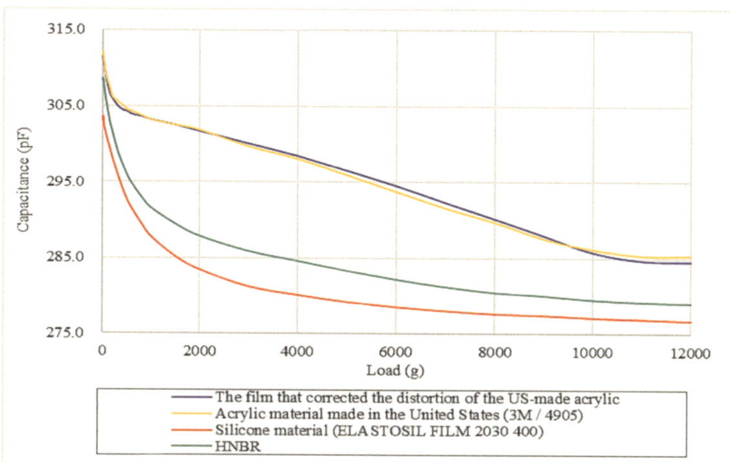

Figure 18.
Changes in electrostatic capacities due to load when using SWCNT electrodes.

Figure 19 shows the results of SS curves and the results of viscoelasticity tests (storage and loss modulus) of four different elastomers. **Table 2** shows the hardness of the elastomer films and the hardness of the resin blocks used in the experiment.

4.1.1 Experiment conducted without resin cushion material

Figure 20 shows the results of measuring the capacitance by increasing the load from 10 kg to 120 kg at 10 kg intervals. Furthermore, in **Figure 20**, "Change in electrostatic capacity due to load" shows the measured values from 10 kgf to 120 kgf, but this is a value limited by the measurement range of the 120 kg load cell used for the measurement. It has been confirmed that the measurement range of the sheet-type DE sensor used in this experiment can be measured from about 4 kgf (using acrylic

Material	Measurement range (g)		
a) Measurement range when carbon grease is used for the electrodes.			
Silicone material (ELASTOSIL FILM 2030 250)	30 ~ 9,000		
Modified HNBR	20 ~ 8,500		
Acrylic material made in the United States (3 M / 4905)	10 ~ 8,000		
The film that corrected the distortion of the US-made acrylic	8 ~ 8,000		
b) Measurement range when carbon black is used for the electrodes.			
Silicone material (ELASTOSIL FILM 2030 250)	30 ~ 10,000		
Modified HNBR	20 ~ 9,000		
Acrylic material made in the United States (3 M / 4905)	10 ~ 8,500		
The film that corrected the distortion of the US-made acrylic	8 ~ 8,500		
c) Measurement range when SWCNT is used for the electrode.			
Silicone material (ELASTOSIL FILM 2030 250)	30 ~ 11,000		
Modified HNBR	20 ~ 10,500		
Acrylic material made in the United States (3 M / 4905)	8 ~ 10,100		
The film that corrected the distortion of the US-made acrylic	5 ~ 10,100		
Material	**Maximum measurement speed (msec)**		
	Carbon grease	**Carbon black**	**SWCNT**
d) Electrodes and measurement speeds.			
Silicone material (ELASTOSIL FILM 2030 250)	69	66	59
Modified HNBR	80	78	68
Acrylic material made in the United States (3 M / 4905)	93	86	73
The film that corrected the distortion of the US-made acrylic	94	88	74
Note: Each of the above measurement data was confirmed in the load cell.			

Table 1.
Pressure measurement results of donut-type sheet DE sensor.

film with corrected distortion) by changing the load cell used for measurement. However, in order to unify the experimental conditions, only the values measured with a 120 kg load cell are shown in **Figure 20**.

It can be seen that the capacitance of acrylic changes almost linearly from 10 kg to the measurement limit of 120 kg. On the other hand, silicon changed linearly from 30 kg, but the amount of change decreased from 100 kg. In the case of silicon, the change in capacitance stabilizes in about 10 seconds, but in the case of acrylic, it took about 20–25 seconds. Since the silicon used in this experiment is harder than acrylic, the region that changes linearly is narrow, but the response speed is considered to be fast. HNBR had intermediate values between silicon and acrylic.

4.1.2 Demonstration experiment to see if bolts are tightened evenly

As explained above, capacitance was measured while tightening only the lower right bolt (No. 1) (see **Figure 12**) by 1/2 or 1 turn compared with the others. As a

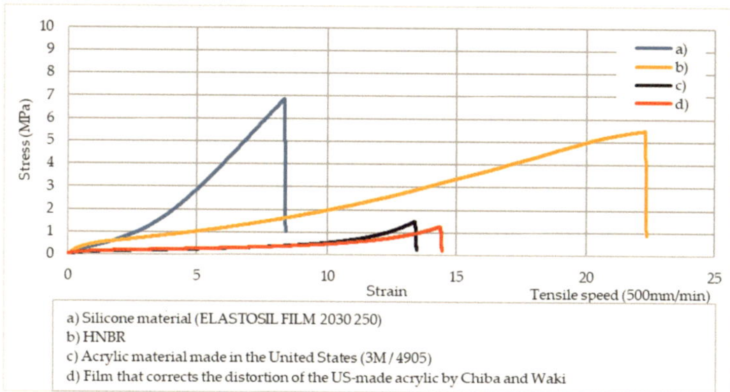

(a)

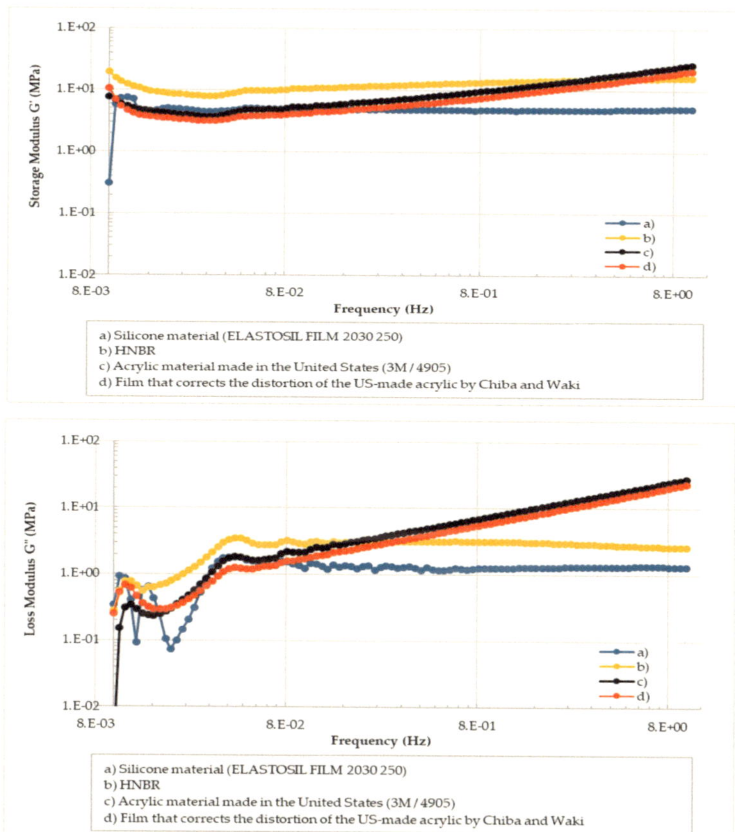

(b)

Figure 19.
The results of SS curves and the results of viscoelasticity tests (storage and loss modulus) of 4 different elastomers. (a) Tensile test results for 4 types of elastomers. (b) 4 types of Dynamic viscoelasticity test results (upper figure: Storage, lower figure: Loss modulus).

result, it was confirmed that the pressure near the bolt at the lower right was higher than the others. **Figure 21** is a graph showing the change in capacitance when bolts are tightened. When only the lower right screw (CH1) was rotated 1/2, the capacitance

Material	Hardness (Asker C)
a) Elastomers used for DE sensors	
HNBR	50
Acrylic material made in the United States (3 M / 4905)	18.0
The film that corrected the distortion of the US-made acrylic	17.5
Silicone material (ELASTOSIL FILM 2030 400)	59.0
b) Resin blocks.	
Material	Hardness (Asker C)
Acrylic gel sheet (Blue Forest Trading LLC:KG-01)	10
Urethane gel (EXSEAL Co., Ltd.:H5–100)	2

Note: This time, the experiment was conducted using only the acrylic gel sheet. Next time, we plan to perform comparative verification using a softer urethane gel sheet.

Table 2.
The hardness of the elastomer films and the hardness of the resin blocks used in the experiment.

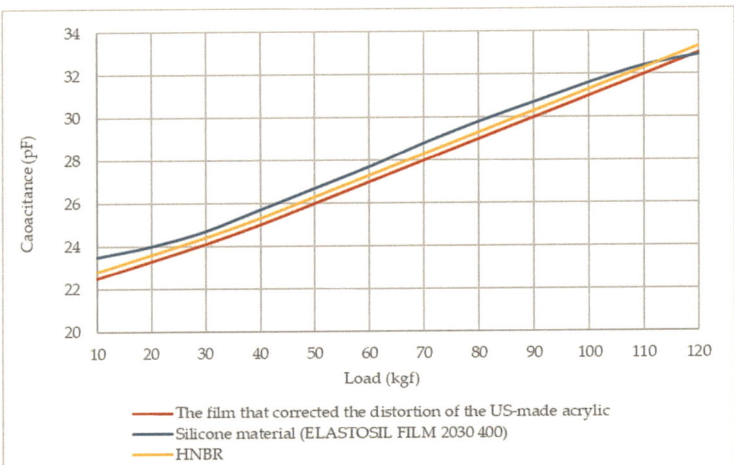

Figure 20.
Change in capacitance due to load.

was about 22.42 pF. This is equivalent to 8.46 kg when converted to load. In addition, the capacitance when only the lower right screw is rotated once is about 24.15 pF, which is equivalent to 30.89 kg.

4.2 Stretch sensor elongation comparison due to differences in the performance of different types of electrodes

Figure 22 shows the change in conductivity when a DE sensor with carbon grease, carbon black, and SWCNT electrodes attached to the acrylic that has been subjected to

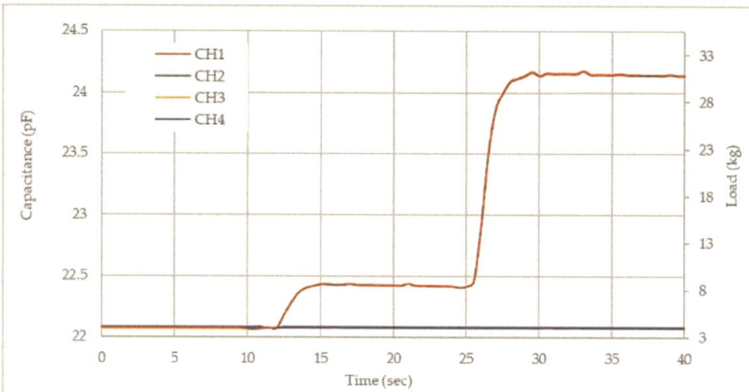

Figure 21.
*Change in capacitance due to measured load: * the load was calculated from the capacitance based on* ***Figure 13****.*
** CH1: Lower right, CH2: Upper right, CH3: Upper left, CH4: Lower left.*

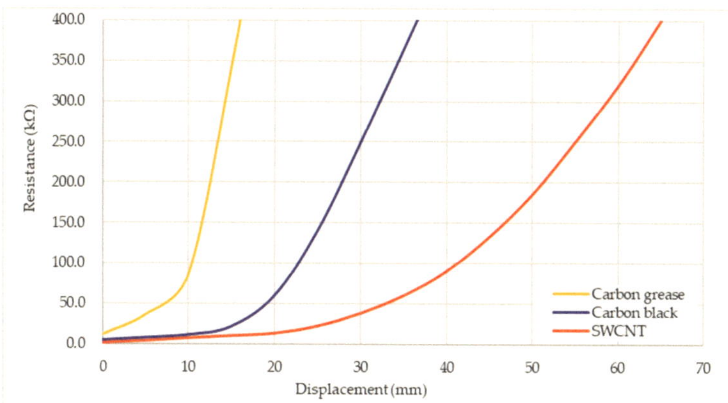

Figure 22.
The change in conductivity when a DE sensor with carbon grease, carbon black, and SWCNT electrodes attached to the acrylic that has been subjected to strain removal treatment is stretched by 5 mm.

strain removal treatment is stretched by 5 mm. As mentioned in above, the thickness of the carbon grease electrode was 100 μm.

The thickness of the carbon black was 70 μm and that of the SWCNT was 50 μm.

5. Discussion

Section 5.1 discusses the results of the experiment in this time, and Section 5.2 discusses the possibility of applying DE sensors to robotics and human-robot interaction (HRI).

5.1 Discussion of experimental results

First, the values obtained with all the data in **Table 1a**, **b** and **c** and the sensors with the four SWCNT electrodes used in the flanges were compared with the values obtained with the load cell in each case. The values of both types of sensors were almost the same. With the donut-shaped DE sensor, there was no significant difference in the measured pressure range due to the difference in the electrodes. The case where SWCNT was used for the electrode had the widest measurement range. The reason for this will be described in the stretch sensor section below. When the SWCNT electrode was used for the acrylic from which the strain had been removed, a change in capacitance appeared from the time the load reached 5 g, and a change in capacitance was confirmed up to 10,100 g. With this electrode, however, it is considered that the change in capacitance was small because the deformation of the resin block reached the limit at 10,000 g or more. With this electrode, it is considered that the change in capacitance was small because the deformation of the resin block reached the limit at 10,000 g or more. Moreover, it changes almost linearly between 500 g and 10,000 g. In the case of the sensors used the other electrodes, however, the boundary that changes linearly is narrower than that. In this sensor, the capacitance changes by about 5% by changing the test frequency from 120 Hz to 1 kHz. The reason is considered to be the influence of the frequency characteristics of carbon grease. When this electrode is used, the frequency characteristic of AC is worse than expected, and the capacitance changes more than the deformation of the elastomer.

We also conducted a demonstration experiment to see if it was possible to detect abnormal bolt tightening on the flange. After tightening the bolt in the lower right of the flange more than the other three places, we measured the pressures in four places (including where the bolt was tightened more) with the load cells and DE sensors using SWCNT electrodes, and each was compared. As a result, it was confirmed that the pressure of the tightened bolt was higher than the others. In the experiment above, the DE sensor was placed on the left side of the bolt, but for confirmation, the DE sensor was placed again on the right side of the bolt, the pressure was checked, and it was confirmed that the values were the same. When actually using these sensors, it is better to make a hole in the DE sensor and put it though in the bolt. It seems that this can be applied to various joints in the future, for example, by tightening pipes and tires that allow dangerous substances to flow, to support safer operation. The reason why the capacitance was measured by tightening the lower right bolt (No. 1) with two steps (1/2 or one turn) is that if those were suddenly tightened with a large force, they might be damaged. Therefore, two-stage measurement was employed.

In the case of the DE sensor without cushion, in the preliminary study, the film thickness of the silicon was as thin as 0.5 mm, and it was quite hard, so the range of change was estimated to be narrow. However, even with silicon, it was confirmed that the load changed from 10 kg to 100 kg linearly from 30 kg to 100 kg. The sheet-type sensor using acrylic shows a change in capacitance from a load of several kg, and changes almost linearly from 10 kg to the measurement limit of 120 kg. HNBR showed an intermediate value between them. In any case, with a load of 10 kg or less, linear displacement did not occur, and with acrylic, the limit was about 4 kg. The silicon weighed 9 kg and the HNBR weighed around 7.5 kg. When measuring a load lower than this value, it seems possible to measure it sufficiently by adding a cushion material (this discussion will be described in detail below).

When measuring cases with large pressures, the above three types of DE sensors could measure in a relatively wide range and could be effective in various applications. For example, as mentioned above, these sensors have a thin elastomer of 0.5 mm or less and seem to be effective as a sensor for checking the tightened condition of the screws in the flange part, which is often used for pipelines and plant piping. It may also be used as a sensor to detect failures in the flange due to aging or vibration. The silicon might be better than the acrylic or HNBR for applications that require faster measurement speeds.

Next, looking at the four elastomer materials individually, from the SS curve measurement results (**Figure 19a**) and (**Table 1a** and **b**), the silicon material (ELASTOSIL FILM 2030250) is harder than other films. So, the film seems to have a relatively large measurable pressure change range. That is, there is a drag against pressure, and the film cannot be easily crushed by pressure. The measurement results of dynamic viscoelasticity support this (**Figure 19b**) [26]. In addition, the hardness of the four types of elastomers above was measured using Askar C, and the results were in the order of silicon, HNBR, acrylic with strain, and acrylic with strain removed. Thus, evidences support that the explanation above is correct (See **Table 2**).

As acrylic deforms greatly, it can be detected from the smallest value and can be measured from 5 g. In other words, acrylic is so soft that it can quickly detect even the slightest pressure.

That is, as described in eq. (1), when the elastomer is crushed by pressure or some forces and the electrodes are closer to each other, a larger capacitance can be obtained. On the other hand, with the hardest silicon, no change was observed at about several g, and a force of about 30 g was required.

The difference in electrodes did not significantly affect the measurement range, but in terms of measurement speed, silicon was slightly faster than other films. The reason is that the silicon film is harder than others, so even if it is pushed by pressure, it returns quickly to its original state [26]. The speed of acrylic is a little slower than that of silicon, but it seems to be sufficient for practical use. As shown in **Table 1c**, SWCNTs with good conductivity showed slightly better results. The reason for this is the same as above, and the better the conductivity, the larger the Coulomb force that can be generated, and as a result, the film is distorted quickly, and the reaction is quick to return. There was little difference in measurement speed depending on the electrodes, but it seems that even a small difference in speed could be a key factor for sensors used in smaller devices such as mm size.

By using a very soft resin cushioning material installed in the center of the donut-shaped sheet DE sensor used, it was possible to measure even with a slight pressure, which led to the success of this experiment. Considering the hardness and elongation of the abovementioned elastomer, it can be inferred that the measurement range can be changed by changing the material of the cushioning material installed on the elastomer film [29]. As mentioned in the Background of dielectric elastomers (DEs) above, if the shape is semicircular, it could come into contact with the object to be measured faster. In such a form, as the pressure or pushing force increases, the ground plane of the elastomer becomes larger by that amount, and it is considered that a sufficient effect in terms of capacitance is produced. In this experiment, a small cushion was sufficient without having to make such a semicircular sensor shape. In this experiment, we did not conduct an experiment to change the shape of the cushion material. If the shape of the cushion material is such that the convex lens is viewed from the side, it is considered that the pressure range width can be increased while maintaining the small shape.

Furthermore, we will conduct the experiment again using the ultra-soft modeling resin: H5-100/H5-600J] manufactured by EXCEL Co., Ltd., as the new cushioning material. This material is softer than the resin used this time, so it seems that it would be possible to detect even with a smaller pressure (see **Table 2b**). The minimum measurement depends on the hardness of the elastomer used in the donut-shaped sheet DE sensor, so it is necessary to use a cushion to stretch the elastomer as described above. However, in this respect, acrylic, which has a small minimum measurement value, could be an effective material.

As an interesting result, the pressure measurement curve shape from the sheet type and the donut-shaped DE sensor are opposite of each other. The reason is that in the case of a donut-shaped DE sensor, the elastomer is stretched by the resin block, and then an external force is applied, the resin block shrinks, the elongation of the elastomer is attenuated, and the capacitance is reduced (See **Figures 16–18**). In the sheet type, an external force is directly applied to the sheet, and the entire sheet is stretched, whereby the amount of electrostatic charge increases (see **Figure 19a**). The reason for conducting the donut-type and sheet-type experiments was that we wanted to prove that the donut type could handle the pressure of the gram order to the 10 kg order, and the sheet type could handle the pressure of the 10 kg order to the 100 kg order. As shown in the results of this experiment, this is because that it might be able to serve as a guideline for proper development within industries.

As shown in **Figure 19a**, one more interesting finding is that when comparing the SS curves of a film from which the strain of the film has been removed with 3 M acrylic and the SS curve of the film without removing the strain, the curves are almost the same to some extent, and the breaking stress is also almost the same. However, it turned out that the final elongation was different. This is thought to have caused a difference in the pressure measurement range and measurement speed in this experiment. It is presumed that the film was more uniform and therefore more easily stretched (see **Figure 19a**). In addition, since 100% pre-stretching was applied, the hardness of the acrylic became moderately hard, and dynamic viscoelasticity was also moderately present (see **Figure 19b**).

The various DE sensors in this experiment can measure any pressure changes from deformation instantaneously within the measurement range. It might depend more on the circuit design. The circuit used this experiment was the same to the circuit mentioned in Section 3.3. If a circuit having a higher degree of amplification is used for this circuit, even a smaller change can be measured, so that the measurement range can be expanded. As mentioned above, the selection of the elastomer material and cushioning material used for the sensor is important for improving the measurement speed, but the tune-up of the circuit is also important for achieving a faster measurement speed.

In 2011, the authors of this paper, in collaboration with the Japan Agency for Marine-Earth Science and Technology (JAMSTEC), examined a DEA that can be driven even on a 100,00 m deep seabed [30, 31]. Using a pressure chamber that could reproduce the pressure of the seabed of 10,000 m, a roll-type DEA was manufactured by rolling a sheet-type DE using an acrylic material, and when an experiment was conducted, it was able to be sufficiently driven even under a pressure of 10,000 m. We believe that the DE sensor can be sufficiently driven as a sensor even under a pressure of 10,000 m if an elastomer with sufficient thickness is used and a cushion material with an appropriate thickness is selected.

Next, regarding the stretch sensor, the silicon material is harder than other films, and as a sensor, the operation speed is fast. However, since the elongation width is small, it seems that the intended use could be limited. Materials with low dynamic viscoelasticity, such as silicone, might not be well suited for DEA/sensor materials [26]. In comparison, acrylic is softer and more stretchable than other films. It could be suitable for DEA/sensor materials. The reaction speed is also sufficient for practical use, and it seems to be optimal as a sensor that matches human movements. In addition, the pressure change range that can be measured is large.

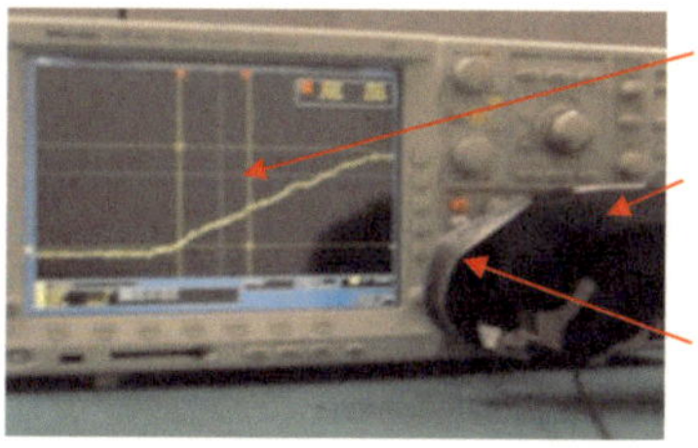

Figure 23.
Rubber glove with DE sheet sensor.

Acrylic's upward-sloping dynamic viscoelasticity supports this evidence (see **Figure 19b**).

The difference in conductivity between carbon grease, carbon black, and SWCNT electrodes is clear, and SWCNT is the most suitable for stretch sensors (see **Figure 22**). The SWCNTs are more conductive, but not only that, SWCNTs are packed in a spray and sprayed to achieve a uniform, thinner electrode [25]. If you look at a commercial basis, however, the cost of carbon black is much cheaper than SWCNTs and multi-walled CNTs (MWCNTs). So, this choice should not be disregarded.

In 2012, we created a system to check how much a finger bends by attaching a DE sensor with a carbon black electrode on a non-distorted acrylic film to a rubber glove. (See **Figure 23**) [31–33].

However, due to the carbon black, the stretch sensor that follows the movement of the finger did not stretch sufficiently. So, we would like to verify it again with a system that reflects these data.

5.2 Application to robotics and human robot interaction (HRI)

The best use of the DE pressure sensor is to apply it to finger pressure sensors in robots or magic hands. This type of sensor is thought to be able to measure pressure ranging from 1 gram to about 150 kg. Additionally, it is extremely flexible, thin, and small, so dexterity rivaling that of human fingers could be realized. When working with humans, it goes without saying that a robot with humans-like abilities would be easier to work with. S. Chiba and M. Waki used a DEA to create a model in which a robot finger is driven by tendons similar to those found in a human finger. Next time, we plan to incorporate DE sensors into this model to create a finger that is more like a human finger and has senses [34].

If this stretch sensor is used well, it can be used as a three-dimensional position sensor for the arms and legs of robots and the like. The principle of 3D position sensor will be explained step by step below.

A stretch sensor is attached to the upper and lower sides of the robot's arm to move the arm upward. The upper sensor is extended and the lower sensor is the contracted. As a result, it is possible to two-dimensionally determine at what angle the arm is bent. Furthermore, by deploying another pair of sensors on the side of the arm, it is possible to sense diagonal movement. That is, the three-dimensional position can be easily determined from the difference between the upper and lower capacitance and the difference in the capacitance on the side surfaces. For this, a calculation table might be helpful. In the next experiment, we plan to attach such a sensor to the robot. As it is now possible to lift an 8 kgf weight by 1 mm or more at 88mms with 0.15 g (0.96 g including reinforcement material) of acrylic [25, 35, 36], we will use it for this system. If you created a finger model that incorporates DEA and DE sensors, it seems that you could make a human-like finger. In addition, it seems that the arm with three pairs of the above sensors could detect the speed of movement of the arm. The expansion and contraction speed of DEA moving in the three-dimensional direction can be calculated from the change in capacitance and the position of the arm (based on the information from the position sensor). Alternatively, each value could be imported as a variable into such a calculated table. Environmental resistance is also an important factor for the actual use of robots. Regarding the environmental resistance of DE sensors, silicon with SWCNT electrodes and HNBR sensors can be used at temperatures 150–200°C [37]. At low temperatures, silicon is said to be usable down to around −100°C [38]. If it could be used in this temperature range, it might be expected to be used as a robot sensor for space environments, deep seas, and extremely cold regions.

6. Conclusion

The difference in the type of elastomers, the thicknesses and shapes of the cushioning materials placed under the elastomer films, the difference in the presence or absence of distortion of the films, the effect of the difference in the type of electrodes were examined, and the following conclusions are obtained:

- The silicon sensor has a small deformation, but the reaction speed as a sensor is fast.

- The acrylic sensor has a large amount of deformation, and the reaction speed as a sensor is practically sufficient. Therefore, it is suitable for stretch sensors. In addition, as a pressure sensor, it can measure pressure in a certain range width steplessly.

- In acrylic, compared with strain-removed films and the untreated films, the treated films had better performance.

- Compared with the difference in measurement speed between each electrode, but there was no significant difference.

- When four acrylic sensors using SWCNT as electrodes were installed and a bolt was tightened in a certain place more than the other three, the pressure was higher than the other three.

- The cushioning material allows you to extend the measurement pressure range. However, even in the case without cushioning materials, the sensors using the SWCNT electrodes were able to measure the pressure accurately in a large pressure range of about 10 kg or more.

Acknowledgements

We would like to thank Mr. M. Uejima, Mr. H. Uchida, and Mr. M. Takeshita of ZEON Corporation for providing SWCNT (ZEONANO® -SG101) and HNBR free of charge for carrying out our experiment.

Author details

Seiki Chiba[1]* and Mikio Waki[2]

1 Chiba Science Institute, Tokyo, Japan

2 Wits Inc., Sakura, Tochigi, Japan

*Address all correspondence to: epam@hyperdrive-web.com

References

[1] Ichinose S. Current status and future of new materials for sensors. The Robotics Society of Japan. 1991;**9**(7): 883-887

[2] Sarutani T, Takahashi K, Aga T, Yamagata T. Small integrated pressure sensor. The Institute of Electronics, Information and Communication Engineering. 1991;**J74-C2**(5):333-339

[3] Shirakawa H. Discovery of conductive polymers and conductive mechanism. Journal of Chemistry and Education. 2019, The Chemical Society of Japan;**67**(2):82-85. DOI: 10.20665/kakyoshi.67.2_82

[4] Harada Y, Suehara T, Ueno T, Higuchi T. Development of an actuator using thermal deformation. In: Proceedings of the Spring Meeting of the Japan Society for Precision Engineering. Tokyo: The Japan Society for Precision Engineering; 2006. pp. 803-804

[5] Piezoelectric, CEGA Corp. Available from: cega.jp/electronic/piezoelectric/

[6] Otaka H. Development of stretchable strain sensor system. Journal of The Society of Rubber Science and Technology. 2018;**91**(2):41-48

[7] Pelrine R, Chiba S. Review of artificial muscle approaches. In: Proceedings of the Third International Symposium on Micromachine and Human Science, Nagoya, Japan. 1992. pp 1-9

[8] Pelrine R, Kornbluh R, Joseph J, Chiba S. Electrostriction of polymer films for microactuators. In: Proc. IEEE Tenth Annual International Workshop on Micro Electro Mechanical Systems. Nagoya, Aichi, Japan: IEEE; 2009. pp. 238-243. DOI: 10.1109/MEMSYS.1997.581728

[9] Chiba S, Waki M. Evolving artificial muscles-further expanding applied technology. In: Proc. of the 27th Annual Conference of the Robotics Society of Japan. Yokohama, Kanagawa, Japan: Yokohama National University; 2009

[10] Sun H, Liu X, Liu S, Yu B, Ning N, Tian M, et al. A supramolecular silicone dielectric elastomer with a high dielectric constant and fast and highly efficient self-healing under mild conditions. Journal of Materials Chemistry A. 2020; **8**:23330-23343. DOI: 10.1039/D0TA06577C

[11] O'Brien B, Thode J, Anderson I, Calius E, Haemmerle E, Xie S. Integrated extension sensor based on resistance and voltage measurement for a dielectric elastomer. In: Proceedings, SPIE Smart Structure and Material + Nondestructive Evaluation and Health Monitoring, 652415. San Diego, California: UAS; 2007. pp. 23330-23343. DOI: 10.1117/12.715823

[12] Han M, Lee J, Kim J, An H, Won Kang S, Jung D. Highly sensitive and flexible wearable pressure sensor with dielectric elastomer and carbon nanotube electrodes. Sensors and Actuators A: Physical. 2020;**305**:111941. DOI: 10.1016/j.sna.2020.111941

[13] Liebscher H, Tahir M, Wiessner S, Gerlach G. Effect of barium titanate particle filler on the performance of polyurethane-based dielectric elastomer actuators. In: Proceedings, SPIE Smart Structure and Material + Nondestructive Evaluation, 1204210. Long Beach, California: UAS; 2022. DOI: 10.1117/12.2612354

[14] Bae J, Chang S. PVDF-based ferroelectric polymers and dielectric elastomers for sensor and actuator

applications: A review. Functional Composites and Structures. 2019;**1**(1):29. DOI: 10.1088/2631-6331/ab0f48

[15] Briggs C, Kaiser G, Sporidis Y, Vicars P, Rasmussen L, Bowers M, et al. Sensitive and robust electroactive polymer tactile pressure sensors and shape-morphing actuation for robotic grippers. In: Proceedings, SPIE Smart Structure and Material + Nondestructive Evaluation, 1204201. Long Beach, California: UAS; 2022. DOI: 10.1117/12.2607779

[16] Agostini L, Monari E, Caselli M. Multidirectional hemispherical dielectric elastomer proximity sensor for collision avoidance in human-robot interaction applications. In: Proc. SPIE 12042, Electroactive Polymer Actuators and Devices (EAPAD) XXIV, 1204202. Long Beach, California, USA: SPIE; 2022. DOI: 10.1117/12.2612864

[17] Jung K, Kim K, Choi H. A self-sensing dielectric elastomer actuator. Sensors and Actuators A: Physical. 2008; **143**(2):343-351. DOI: 10.1016/j.sna.2007.10.076

[18] Bose H, Fub E. Novel dielectric elastomer sensors for compression load detection. In: Proceedings, SPIE Smart Structure and Material + Nondestructive Evaluation and Health Monitoring, 905614. San Diego, California: UAS; 2014. p. 13. DOI: 10.1117/12.204513

[19] Rosenthal M, Bonwit N, Duncheon C, Heim J. Application of dielectric elastomer EPAM sensors. In: Proceedings, SPIE Smart Structure and Material + Nondestructive Evaluation and Health Monitoring, 65241F. San Diego, California: UAS; 2007. p. 7. DOI: 10.1117/12.715048

[20] Walker C, Andersen I. Monitoring diver kinematics with dielectric elastomer sensors. In: Proceedings, SPIE Smart Structure and Material + Nondestructive Evaluation and Health Monitoring, 1016307. Portland, Oregon: UAS; 2017. DOI: 10.1117/12.2260394

[21] Larson C, Spjut J, Knepper R, Shepherd R. A deformable Interface for human touch recognition using stretchable carbon nanotube dielectric elastomer sensors and deep neural networks, soft. Robotics. 2019;**6**(50): 611-620. DOI: 10.1089/soro.2018.0086

[22] Chiba S, Kornbluh R, Pelrine R, Waki M. Artificial muscle and their next generation. In: Proc. International Symposium on Organic and Inorganic Electronic Materials and Related Nanotechnologies. Nagano, Nagano Pre., Japan: The Chemical Society of Japan; 2007

[23] Venkatraman R, Kaaya T, Tchipoque H, Cluff K, Asmatulu R, Amick R, et al. Design, fabrication, and characterization of dielectric elastomer actuator enabled cuff compression device. In: Proc. SPIE 12042, Electroactive Polymer Actuators and Devices (EAPAD) XXIV, 1204205. Long Beach, California, USA: SPIE; 2022. DOI: 10.1117/12.2613250

[24] Chiba S. Electroactive polymer artificial muscle (EPAM) in human life science. In: Proc. of Human Life Science Forum (Invited). Japan: Kyoto University; 2005

[25] Chiba S, Waki M, Takeshita M, Yoshizawa T. Improvement measures for components of dielectric elastomers for heavy duty uses such as robots and power assist devices. Advances in Theoretical and Computational Physics. 2021;**4**(3):241-249. DOI: 10.33140/ATCP.04.03.08

[26] Chiba S, Kobayashi M, Qu T, Zhu S, Waki M, Takeshita M, et al. Examination

of factors to improve the elongation and output of dielectric elastomers. In: Proc. SPIE 12042, Electroactive Polymer Actuators and Devices (EAPAD)XXIV, 1204211. SPIE; 2022. DOI: 10.1117/12.2603716

[27] Chiba S, Waki M, Takeshita M, Uejima M, Arakawa K. Dielectric elastomer using CNT as an electrode. In: Proc. SPIE 11375, Electroactive Polymer Actuators and Devices (EAPAD), XXII, 113751C. 2020. DOI: 10.1117/12.12.2448512

[28] Chiba S, Waki M, Jiang C, Takeshita M, Uejima M, Arakawa K, et al. The possibility of a high-efficiency wave power generation system using dielectric elastomers. Energies. 2021;**14**: 3414. DOI: 10.3390/en14123414

[29] Chiba S, Waki M. Application to dielectric elastomer materials, power assist products, artificial muscle driven. In: Next-Generation Polymer Development New Application Development and Future Prospects. Tokyo, Japan; Section 3, Chapter 4: Technical Information Association; 2019 ISBN-10; 4861047382 ISBN-13: 978-4861047381

[30] Chiba S, Waki M, Sawa T, Yoshida T, Kornbluh R, Pelrine R. 2011 Electroactive polymer artificial muscle operable in ultra-high hydrostatics pressure environment IEEE Sensors Journal. 2011;**11**(1):3-4. ISJEAZ (ISSN 1530-437X). DOI: 10.1109/JSEN.2010.2053702

[31] Chiba S, Waki Fujita K, Song Z, Ohyama K, Zhu S. Recent Progress on soft transducers for sensor networks. In: Technologies and Eco-Innovation towards Sustainability II. Singapore: Springer Nature; 2019. pp. 285-298. DOI: 10.1007/978-981-13-1196-3_23

[32] Chiba S, Waki M, Wada T. Evolving dielectric elastomer artificial muscles. Petrotech. 2012;**35**(7):501-506

[33] Waki M, Chiba S. Elastomer transducers. Soft Actuators. Japan: Springer; 2014. pp. 447-460. DOI: 10.1007/987-4-431-54767-9.

[34] Chiba S et al. "Elastomer Transducers", Advances in Science and Technology. Vol. 97. Switzerland: Trans Tech Publication; 2016, ISSN: 1662-0356. pp. 61-74. DOI: 10.4028/wwwscienctific.net/AST.97.61

[35] Chiba S, Waki M, Ono K, Hatano R, Taniyama Y, Okada E, Ohyama K. Challenge of creating high performance dielectric elastomers, In Proc. of SPIE2021 Smart Structures and Materials Symposium and its 23rd Electroactive Polymer Actuators and Devices (EAPAD) XXIII, 115871T, 2021: pp. 1157-1162. DOI: 10.1117/12.2581255

[36] Chiba S, Waki M, Takeshita M, Yoshizawa T. Application of dielectric elastomers to actuators/robots. Power assist equipment, etc. In: Design of Conductive Materials, Conductivity Controls and Latest Application Developments. Japan: Technical Information Institute Co., Ltd; 2021. pp. 911-920 ISBN:978-4-86104-867-8-C3054

[37] Aimura Y. Basic properties and applications of hydrogenated nitrile rubber. Journal of The Society of Rubber Science and Technology. 1997;**70**(12): 681-688

[38] Taiyo Wire Cloth Co. LTD. Available from: https://www.twc-net.co.jp/products/genre/Adhesives/

Chapter 7

Perspective Chapter: Fabulous Design Speed Industrial Robotic Arm

Falih Salih Mahdi Alkhafaji

Abstract

This chapter focuses on the design of a handling 5 Degree of freedom (DOF) robot arm model for industrial application. Optimal trajectory planning of industrial robots in the assembly line is a key topic to boost productivity in a variety of manufacturing activities. The aim is to improve the speed performance using multi techniques starting from estimating the transfer function of each manipulated joint, then designing the controller for each DOF reached to modeling arm motion. The designed model has been developed the structural design and testing motion characteristics by using SolidWorks and Simscape toolbox. To enhance the speed performance, it is proposed a High-Speed Proportional Integral Derivative controller (HSPID) based on an improved GA. The comparison response time between uncontrolled and controlled systems proves that the proposed controller produced extreme reduction responses to be measured within the Microsecond unit. Based on trajectory motion, the efficiency of the proposed method is assured by case study motions. The innovative design offers the best solution to rise accurate performance and productivity.

Keywords: industrial robot arm, TF, Simscape, trajectory planning, HSPID, response time

1. Introduction

For decades, polymorphic robots are one of the most widely used mechatronic systems in the industry, which has generated the need to constantly create industrial robots with more power, speed, and precision for a wide variety of tasks to improve automation systems precisely offering lower-cost production [1–4]. The manufacturing sector is moving towards industry 4.0 and demands high-end automation in the process [5]. Due to their great usefulness in the industry, industrial robot arms are widely used to move material, parts, and tools as well as the welding and painting of parts [6], offers a powerful performance in terms speed, repetitive and seemingly intelligent decisions [7–10]. Structurally, a robot arm is constructed from several parts such as manipulator links, actuators, controller, and sensors where the controller acts as the brain to manipulate the mechanical parts [11–13]. A robot arm manipulator is constructed from links and joints to control the robot's trajectory. The links of such a manipulator are connected by joints allowing either rotational motion such as in an articulated robot or translational (linear) displacement. Usually, the end effector is attached to the end joint of the structure [14–17]. Precise control at each DOF of a robotic arm is considered a competitive key to improving performance through the implementation of industrial robotic arms [18, 19]. The most remarkable controller's strategies are impedance control, force control, motion control, and hybrid motion-force control [20]. On the one side, optimizing the estimation TF model is the most important criterion to enhance controller's design [21]. On the other side, the basis for developing a control system for a robot manipulator is a feedback loop. Therefore, it is necessary to define input signals such as torque and drive input voltage to achieve the desired operation [22, 23]. For several decades, the Proportional Integral Derivative (PID) algorithm is one of the most widely used in the industry for controlling closed-loop feedback systems, because of the less complexity with the ability to meet desired controller's functions for wide scope plant models [24–26]. It stands for three proportional gains: proportional (Kp), integral (Ki), and derivative (Kd), where they should be jointly tuned to get better performance [27, 28]. Despite having only three parameters, the classical PID controller has been unable to meet the sophisticated requirements [29–31]. The problem with PID has been identified as poor tuning, which means that most of the controllers currently in operation have been poorly tuned. This results in a biased judgment against the PID controllers themselves [32]. The controller tuning greatly affects the control system properties, such as robustness to disturbances and noise [33].

Over 50 years, massive tuning strategies have been suggested to realize satisfied response time characteristics in terms of peak overshoot (Pos), rise time(tr), and settling time (ts). Optimizing PID controller performance can be achieved by considering systematic proportional gain adjustments, otherwise, the tuning will be inadequate and the process testing will take longer [34]. Tuning and optimizing PID gains improves the convergence speed and the global optimization by reduces the overshooting and transition time of the plant system [35]. In fact, an evolutionary algorithm (EA) presents the best solution to optimize PID gains by adapting to the system's nonlinearity [36]. GAs are the first-class category of EA that is commonly used to generate high-quality solutions to optimize PID parameters, by relying on bio-inspired operators such as mutation, crossover, and selection [37, 38]. Usually, GA is applied to optimize a function called fitness function (FF). The FF assists GA algorithms in measuring the quality of the solution of the problem under study and how effective the solution of the problem [39]. In designing the controller, it is related to performance indices such as settling time, integral error, and so on, and might be addressed as a multiobjective function to improve the controller's response [40].

GAs are satisfied in solving multi-objective optimization problems. A possible solution to a problem is considered individual. A group of individuals is called a population. The current population produces a new generation, eventually ceases when it reaches an individual who represents the optimal solution to the problem [41, 42]. A conventional GA requires two variables to be determined, a FF composed of a performance index (*P indices*) and a genetic representation of the solution domain. To formulate the performance index for PID controllers, several related researchers use the following equations to formulate cost functions such as:

integral squared error(ISE); integral absolute error (IAE); integral time squared error (ITSE); integral time absolute error (ITAE); and mean squared error(MSE) as illustrated from Eq. (1) to (5) respectively. Besides that, as it is needed to reduce the error, the FF equation is taken as an inverse of the performance indices as Eq. (6).

$$ISE = \int_0^T |e(t)|^2.d(t) \tag{1}$$

$$IAE = \int_0^T |e(t)|.d(t) \tag{2}$$

$$ITSE = \int_0^T t|e(t)|^2.d(t) \tag{3}$$

$$ITAE = \int_0^T t|e(t)|.d(t) \tag{4}$$

$$MSE = \frac{1}{t}\int_0^T (e(t))^2.d(t) \tag{5}$$

$$FF = \frac{1}{P\ indices} \tag{6}$$

Basically, GA problems rely on three operators: selection, mutation, and crossover [43–48]. In fact, traditional GA generates random population that might produce poor fitness and low-quality of individuals, leading to consume more time to converge through optimized solutions. Therefore, the quality of an initial population of individuals reflected considerably on the GA's performance to produce an optimal solution. Most previous techniques concentrate on the quality of the initial population seeding, such as random initialization, nearest neighbor, and K-means clustering [49–51]. Some researchers used GA to optimize PID, for instance, Guan and Jau [52], Swati K. et al. [53], Tanvir ert et al. [54], Gun B. S. [55], and Apriaskar et al. [56]. On the other side, PSO is one of the EA's that was developed by James Kennedy and Russell Eberhart in 1995, for solving practical issues related to optimization, inspired by the behavior of living things. It has several benefits such as being simple implementation, featuring a simple concept, efficacious computation, and more cost-effective, flexible, and balanced mechanism to improve a global and local exploration abilities [57–60]. Recently, several studies using PSO to optimize PID controllers such as M. I. M. Zakki et al. [61], V. Bagyaveereswaran et al. [62], Y. Xie and J. Meng [63], B. A. Arain et al. [64], S. Howimanporn et al. [65].

Regarding previous works for optimizing PID controllers, the PSO is faster than GA when looking closely at idealistic solutions in case of does not require a detailed mathematical description of the process for formulating the Objective Function (OF) to optimize proportional gains, where the drawbacks of PSO arc the lack of certainty that an optimal solution will be found and even the high computational costs associated with FF [66–68]. Consequently, standard PSOs often fail to solve these complex problems because they easily get stuck in local optima and converge slowly [69]. One of the main differences between PSO and GA is the mechanism of perturbing the solution from the old population to create a new population. These different mechanisms generate a population of solutions with different balances of enrichment and diversification. For GA, the solutions are arranged based on their fitness values [70]. Based on survey, classical GA is not the best solution with respect to PSO. A serious downside of GA comes from the way new generations are computed after the first. It contains a random component that causes generated values to be corrupted during the early stages of global search. Where proposed methods for initial seeding of populations are limited. The limited number of this approach motivates this study because there is room for improvement and finding a better starting population. To improve GA's performance, it was proposed to apply a new technique to GA to raise precise searching constraints by introducing a new Modified Initialization Fitness

Function (MIFF) technique. The proposed was applied to optimize the PID controller for each manipulated joint to enhance velocity performance. Model-Based Design Approach (MBDA) is an advanced simulation technique that is widely used to improve system design, providing explicit models to define activities in the product design and development lifecycle [71]. Additionally, rapid technology using computer-aided design (CAD) enables free-form fabrication of parts with complex geometries directly from CAD models on CNC machines without the need for special fixtures as in the material removal process, provides the best tools for developing products, faster, lower cost and more Competitive Global Market [72, 73]. The collaboration between MathWorks and SolidWorks is one of the best solutions for designing and simulating robot motion, optimizing system parameters, analyzing results in the Simulink environment, analyzing forces due to torque on mechanical joints, provides a nice tool for plotting acceleration due to displacement of arbitrary parts, visualize the motion of CAD assemblies and simplify the physics of mechanical systems without the need to derive equations of motion [74]. The concept and design of robotic arms is not a new concept but still, much work and development are required to enable robots to perform complex tasks. The challenge for robotic technology is to make it compatible with human tasks and hand movements, such as grabbing, swapping, and completing critical tasks. In this way, when we were able to precisely control the movement of the robot, we succeeded in developing the robot arm [75, 76]. In reality, the precise control of each degree of freedom of a robotic arm is a great challenge in implementing industrial work [77]. The robot simulation is used to know the robot torque that will improve and optimize the movement of the arm robot so the robot can help the industry to produce effectively and efficiently [78]. Nowadays, the modeling and control of mechatronic and robotic systems is an open and challenging field of investigation in both industry and academia. The mathematical model of a mechanical system is indeed fundamental for the development of experimental prototypes [79]. On the other side, optimal trajectory planning of industrial robots in the assembly line is a key topic to boost productivity in a variety of manufacturing tasks [80]. The main focus of this work is the design of a 5-DOF industrial robotic arm that further enhanced speed performance and optimized trajectory planning by modeling an HSPID controller based on improved GA (IGA). The GA is implemented based on MIFF as demonstrated in [81]. This provides an opportunity to maximize significantly response time.

The goal is to achieve high-speed performance in designs composed through motion components, which can be concluded as follows: 1) easily create geometric robot parts to assemble; 2) modeling and simulation of the robot arm; 3) Minimize the response time characteristics in terms of tr, ts, and PoS. This can increase the speed of moving the arm from the initial state point to the final state point and significantly reduce the cycle time. SolidWorks was used to design the mechanical structure of the robotic arm. Then perform plugin-based model integration to design controller models within Simscape compatible models, leads to export (XML) files including the part's structure of each component for the base, shoulder, upper and lower arm, and wrist end effector. All the parts are assembled and a HSPID controller is implemented in each DOF of the robot using the Simscape blockset to construct the whole system. Finally, each manipulated joint was examined for in terms of tr, ts, and PoS. This chapter is organized as follows: Section 2 presents the design methodology, including modeling the Simscape configuration, estimating TFs of each manipulated joint, designing the HSPID controller, and a motion detection strategy. Section 3 presents the simulation results approach for estimating TFs, the response time of the designed model with and without a controller, and the simulation results approach for estimating motion based on three different trajectory signals. Section 4 discusses of the results. Section 5 summarizes the results and provides recommendations for future research.

2. Procedure design methodology

This section describes process design techniques including modeling, simulation, and HSPID controllers. Basically, a robotic arm consists of 5 limbs and 5 joints. To verify system-level design and simulation models, we need to define the structure of the system and its behavior. In this work, models related to Solidworks and Simscape were semantically defined using plugin-based model integration to obtain compatible structural and simulatable models leads to export of the models in XML file. The organizational charting procedure design methodology shown in **Figure 1** includes various blocks as shown in the following diagram.

- Mechanical configuration and assembly components.
- Modeling Simscape.
- Estimation TFs of the manipulator's joints.
- Designing HSPID controller for joint's manipulators.
- Mimic trajectory planning.
- Analyze response time characteristics and motion results.

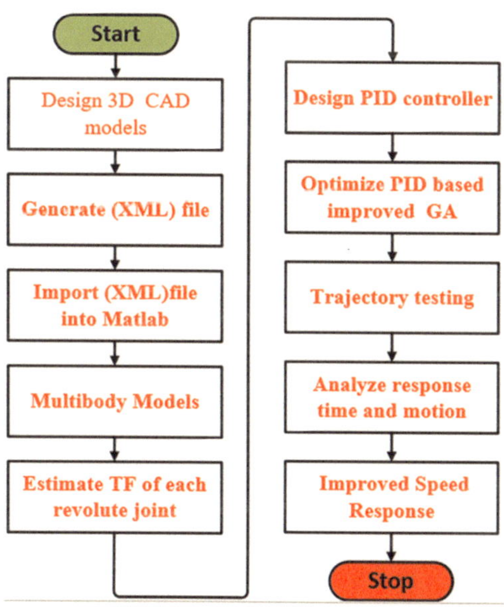

Figure 1.
Scheme of design procedure.

2.1 3D CAD component design and assembly

It is possible to simulate the mechanical behavior of the robot components using axis material during the motion study of the robot, by using SolidWorks in conjunction with Matlab to design high-quality control systems, providing the ability to transform all the component's parameters into XML files, to be imported a dynamic parameter of the physical structure into MATLAB environment, including the inertia matrix calculations for each manipulated joint [82], offers an amazing solution that could be allowed to perform the drawing of a mechanical model to be applied in as a real geometric and mass measurements. In this study, the components (rigid bodies) are built individually and assembled links serially. **Figure 2** presents all the designed rigid bodies of the robotic arm labeled from base to tip: base, shoulder, lower arm, upper arm, connector, Wrist, and end effector. First, we need to attach the base to the robot frame, connect the shoulders to the forearms, connect the upper arms to the forearms via connectors, and connect the wrists to two rigid upper arm bodies. Also, the end effector must be attached to the object for manipulation. In addition to determining the size parameters of the system such as body, height, and weight, the rotation of the rotary joints should be installed. A motion analysis is then presented to verify the effectiveness of the model configuration in terms of velocity and distance. Consequently, different parameters with different settings will lead to various results. **Figure 3** shows the whole system constructed based on 5 joints and 5 links.

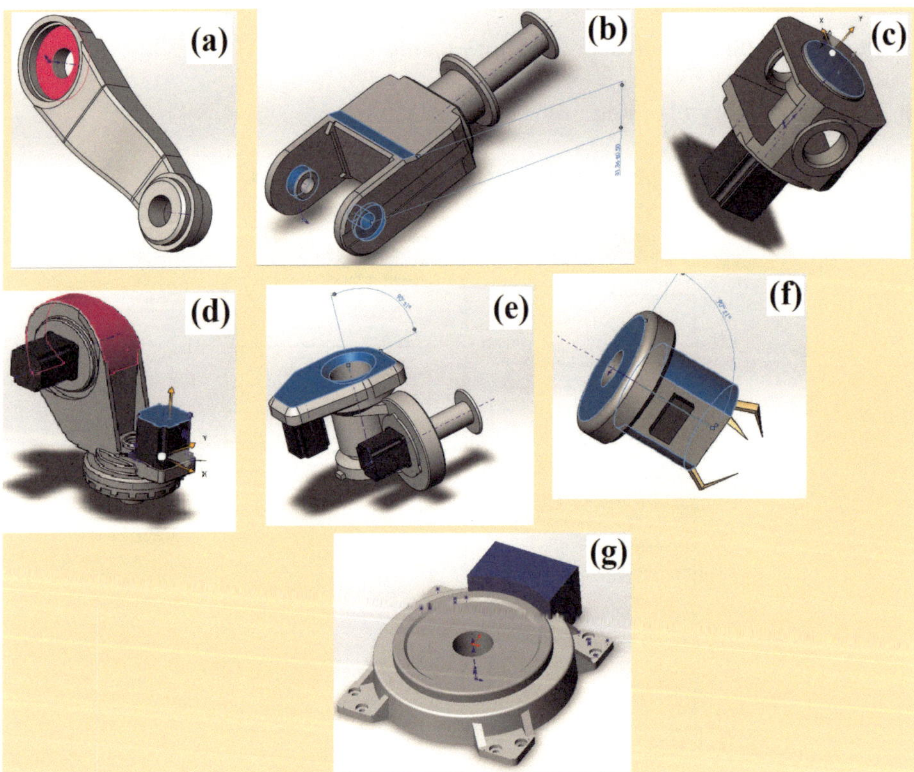

Figure 2.
Three-dimensional solid rigid parts, (a) lower arm, (b) upper arm, (c) connector, (d) shoulder, (e) wrist, (f) end effector, (g) base.

2.2 Modeling Simscape

The Simscape Multibody is used to perform the simulation of links, joints, and dynamic analysis motion. The 3D solid model contains the robot body and a manipulated joint is being used for assembling the whole system. Further, employ SimMechanics tools to produce the XML file for each rigid body. The floor blockset was used as a world frame reference for rigid transformation to set the gravitational force direction. To investigate the motion parameters and response time characteristics, it was used a scope simulator for this purpose. **Figure 4** illustrates the Simscape model of the whole system, representing the joint revolutions and rigid parts as follows; base, shoulder, lower arm, connector, upper arm, wrist, and the end effector.

2.3 Estimate TF of the manipulator's joint

To design a motion controller, we need to estimate the TF for each manipulator joint. This is considered a significant problem in most previous studies, where poor estimation accuracy leads to poor controller design [83]. In this subsection, we present the procedure of estimation TF form utilizing the linear analysis tool for each manipulator's joint, which represented a nonlinearity plant system. Nine steps were performed in the estimation as follows:

- Open the **Model Linearization Tool** from **a** Simscape model.
- **Set Actuation** to **Motion Selection** on the **joint rotation**, and set Sensing on Velocity to make a connection with revolution joint system, then highlight the signals as an analysis point.

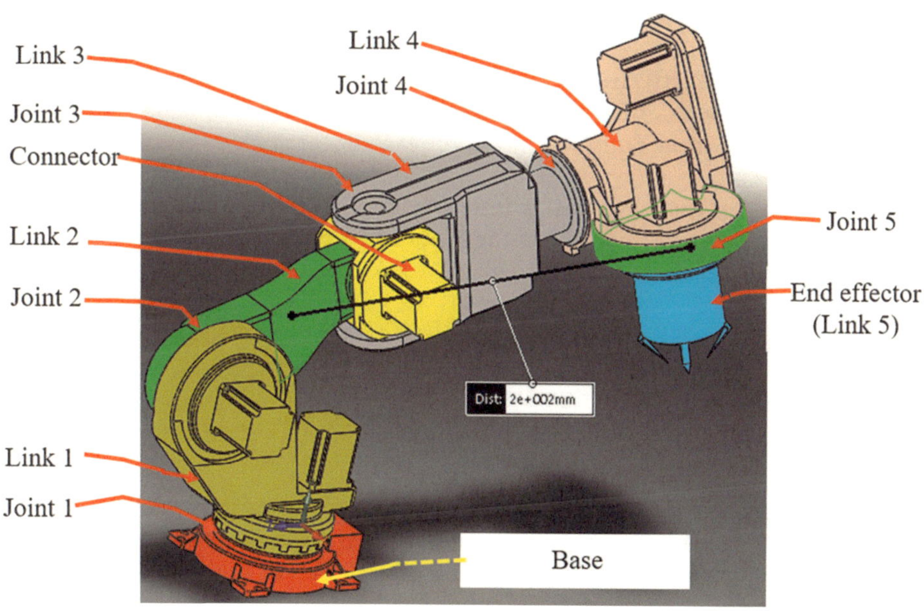

Figure 3.
Designed robot model.

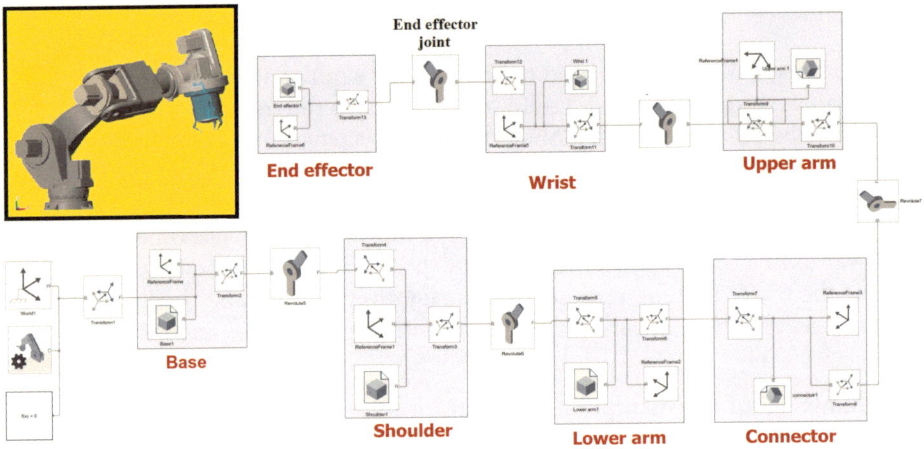

Figure 4.
Simscape model of the whole system.

- Select **Linearization Manager** from the Simulink Apps gallery, to assign the portion of the model to linearize.
- Choose the **Signal** in the model to specify an analysis point for the injected signal. Then select the type of analysis point from the Insert Analysis Points gallery on the Linearization tab.
- Specifying an additive input to a signal by configuring the revolution block's input signal as an **Input Perturbation.**
- Setting the revolution block's output signal to be an open-loop output to take the measurement from the injected signal.
- At this point, the software adds annotations to the model that describe the type of linear analysis point after the analysis points are finished, and identify the type of analysis point for the Simscape model for each linear analysis point.
- Select the linear model (**transfer function**) in the **Linear Analysis Workspace,** then choose **Result Viewer** from the **Plots and Results Viewer** to see the linearized model equations.
- To be defined the TF of the other revolution components, repeat steps 2 and 3 for all signals.

2.4 Proposed IGA

The most important feature of the GA is how it transforms the system output into a fitness value to depress the errors in the reference trajectories of the plant model-based PID controller. Therefore, the chosen IAE evolution or FF should be used to compute the total error between the reference and the system output for each set of generated PID gains. By proposing two techniques the GA could be significantly improved. Firstly, using optimization-based tune (OBT) to find the best optimal solution as explained in [31]. Secondly, initialized FF based on the best optimal constraints to initialize the constraints of GA chromosomes as demonstrated in [73], to modify initialization FF (MIFF) based on the best optimal constraints using CHR tuning method to improve constraint's level of GA chromosomes. This method represented as a CHR_{OBT}, where the proportional gains which resulted by CHR_{OBT} are $\left(Kp_{CHR_{OBT}}\right)$, $(Ki_{CHR_{OBT}})$ and $(Kd_{CHR_{OBT}})$. The mathematical expression of the cost function can be formulated as in Eq. (7).

$$Cost\ Function = \sum_{n=1}^{m} \frac{|rn - yn|}{m} \tag{7}$$

The initial controller gains $Kp(i,j)$, $Ki(i,j)$, $Kd(i,j)$ based on the results of CHR_{OBT} can be modified regarding to generated constraints, which are represented as the following equations:

$$Kp_{MIFF}(i,j) = x1(0) + Kp_{CHROBT} \tag{8}$$

$$Ki_{MIFF}(i,j) = x2(0) + Ki_{CHROBT} \tag{9}$$

$$Kd_{MIFF}(i,j) = x3(0) + Ki_{CHROBT} \tag{10}$$

The population in each generation is represented by a 100 x 4 population (P) matrix as expressed in Eq. (11), to produce chromosomes number in the population. Each row represents one chromosome-based MIFF that compromise $Kp_{MIFF}(i,j)$, $Ki_{MIFF}(i,j)$, $Kid_{MIFF}(i,j)$ and fitness values $FF_{MIFF}(i,j)$ of the corresponding chromosomes.

$$\text{P\_}MIFF = \begin{vmatrix} Kp(1,1)_{MIFF} & Ki(1,2)_{MIFF} & Kd(1,3)_{MIFF} & F(1,4)_{MIFF} \\ Kp(2,1)_{MIFF} & Ki(2,2)_{MIFF} & Kd(2,3)_{MIFF} & F(2,4)_{MIFF} \\ \ldots.. & \ldots. & \ldots. & \ldots. \\ \ldots. & \ldots. & \ldots. & \ldots \\ Kp\ (i,j)_{MIFF} & Ki(\ i,j)_{MIFF} & Kd(\ i,j)_{MIFF} & F(i,j)_{MIFF} \end{vmatrix} \tag{11}$$

The cost function has been minimized subjected to $Kp(i,j)$, $Ki(i,j)$, $Kd(i,j)$ as Eq. (12), Eq. (13), Eq. (14), respectively:

$$Kp\ \min{}_{MIFF} \le Kp(i,j) \le Kp\ max_{MIFF} \tag{12}$$

$$Ki\ \min{}_{MIFF} \le Ki(i,j) \le Ki\ max_{MIFF} \tag{13}$$

$$Kd\ \min{}_{MIFF} \le Kd(i,j) \le Kd\ max_{MIFF} \tag{14}$$

Where $Kp(i,j)$, $Ki(I,j)$, and $Kd(i,j)$ are the optimized proportional gains in the jth area. The error (e_{MIFF}) and the modified cost function-based MIFF might be formulated as in Eq. (15) and Eq. (16), respectively.

$$e_{MIFF} = 1 - \frac{Gp(s) * C(s)}{1 + Gp(s) * C(s)} \tag{15}$$

$$Cost\ function_{MIFF} = \sum_{1}^{n} \frac{|e_{MIFF(n)}|}{m} \tag{16}$$

Where:
(n) is the order of data depending on sampling time.
(m) is the total number of data.

The cost function was written in m-code for each estimated TF, prepared to be imported into GA toolbox, to be run the GA-based MIFF (GA_{MIFF}). The boundaries setting (parameters and operators) were settled as the following:

Iteration: 100; Mutation rate: 0.1; Population size: 100; Arithmetic Crossover; FF: IAE.

2.5 Designing HSPID controller for joint's manipulators

The control topology relies on the injected signals and the feedback position of the end-effector for the demands of the robot application. **Figure 5** shows the Simscape model of the robot arm based on the designed HSPID controller which is constructed from five controllers that are connected with each manipulator's joint, allowing the joints to reach the required angles. It was modeled HSPID controllers regarding each revolution joint IGA to maximize responses with minimal loop interaction and sufficient. The HSPID controller was linked with each revolution joint through motion input with sample time(Ts) of 0.1 s. The proposed HSPID controller model ($C(s)_{MIFF}$) and the plant model (Gp) can be represented in s -domain as Eq. (17) and Eq. (18), respectively.

$$Gp(s) = \frac{\sum_{i=0}^{m} bi\, S^i}{\sum_{j=0}^{n} S^j + aj} = \frac{b0\, S^m + b1\, S^{m-1} + \ldots + bm - 1\, S^m}{aoS^n + a1S^{n-1} + \ldots + an - 1S^n + an} \tag{17}$$

$$C(s)_{MIFF} = Kp_{MIFF} + \frac{Ki_{MIFF}}{s} + Kd_{MIFF} * s \tag{18}$$

2.6 Mimic trajectory planning

The purpose of the robot controller is to send control signals to the joints so that the robot follows a particular path [84]. A trajectory is the robot's position as a

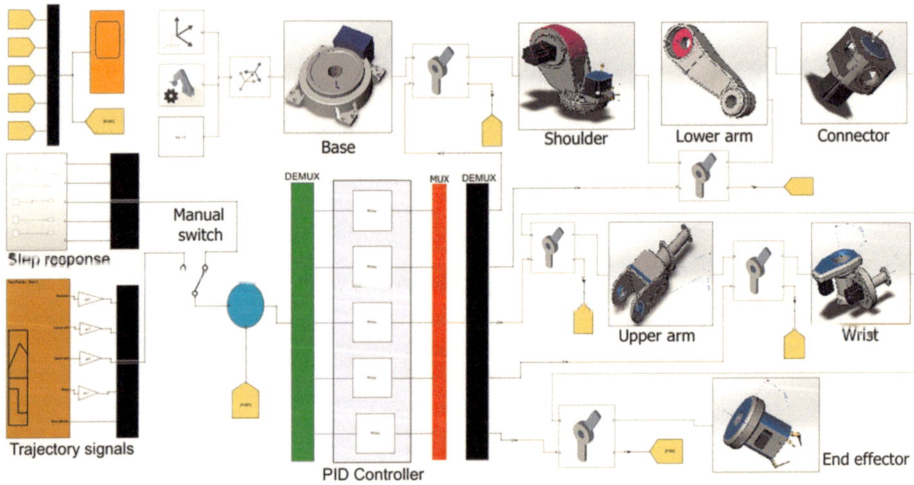

Figure 5.
The proposed HSPID controller based on the whole system.

function of time, and a path is a geometric description of the motion. Simulation software can be used to examine the motion of the arm robot and confirm that the robot follows the path [10, 85]. A trajectory planning of a robot allows us to determine the continuous position paths that will guide the end-effector of the robot [86]. The robot arm is designed for various applications, among which are those where the end-effector to reduce risks industrial risks, in which case a position control is required. The position control of the robot can be approached in two ways, one referring to the joint space and the other to the task space. To verify the trajectory planning, it was utilized Simscape model to simulate the motion while it is in operation bypassing the exact rigid parameters of the whole system into Matlab environment, besides building the trajectory block signals containing a position for five different signals, that the

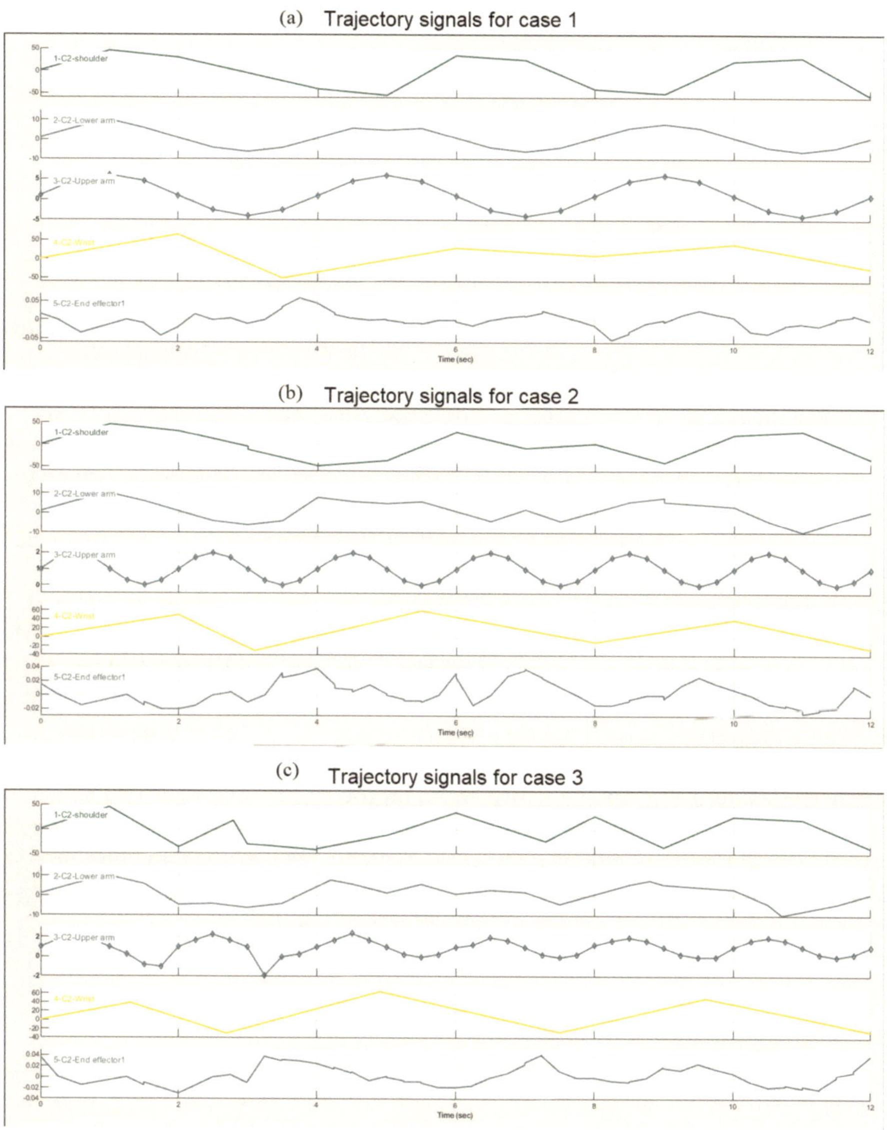

Figure 6.
Trajectory signals under three cases, (a) case1, (b) case 2, (c) case 3.

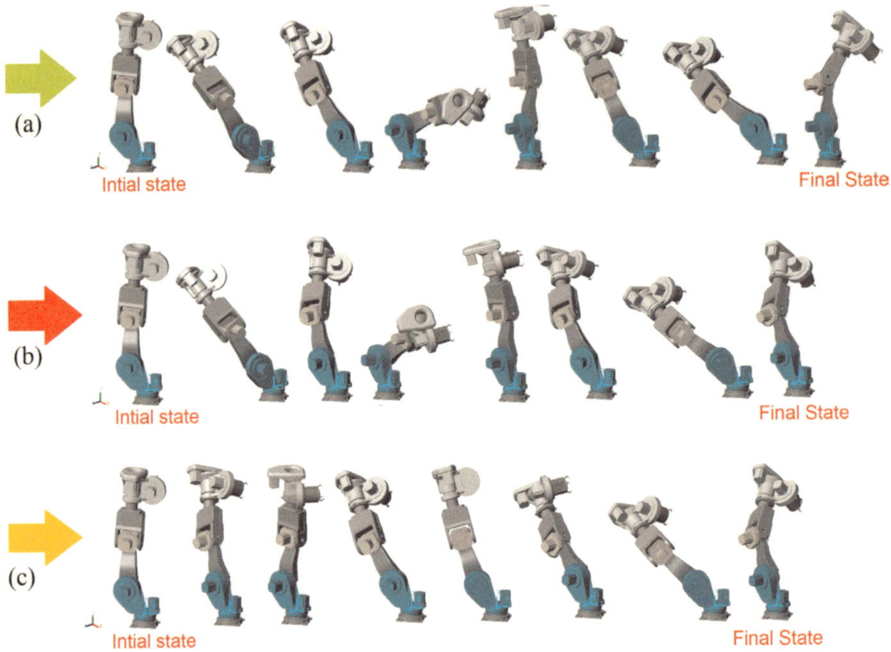

Figure 7.
Robot arm motion simulation-based HSPID controller under three case studies, (a)case 1, (b) case 2, (c) case 3.

robot arm should follow during the motion time. All of the above signals use one of three motion types: Point-To-Point (PTP), in which the robot positions of the trajectory move between them, or Continuous Path (CP), in which the signals move the manipulator tip along a given trajectory and the robot then accurately reproduces it. For displacements, the joint varies from −90° to 90°, and the displacement speed is modified by manipulating the angular frequency at 1 (rad/s).

As shown in **Figure 6**, we prepare three trajectory signals to execute three motion cases over a 12-second time trial to evaluate the maximum admissible torque at the end effector joint (joint 5), beginning from the initial state to the final state as presented in **Figure 7a–c**. The trajectory planning is inspected through multi-shape signals including straight lines, circles, and parabolic curves, according to the robot's sequence to simulate its movement. Therefore, we will run the Simscape model to simulate the model's angular trajectory based on the HSPID and display a 3D animation. Where the joint angles start in a fully extended vertical position at 0 degrees except the shoulder joint angle is fixed at 90 degrees. Finally, the end configuration was determined by the angular positions: shoulder = 60 deg., lower arm = 80 deg., upper arm = 60 deg., Wrist - 90 deg., end effector = 90 deg.

3. Simulation results

In robot simulation, system analysis needs to be done [87]. The simulation results covered four subsections: manipulator joint estimation TF, optimization controller gain, motion results, and response time comparison between proposed controller and without controller.

3.1 Estimation TF of the manipulator joint

Figure 8 illustrates the best estimation TFs for the manipulator's joints for the following components: shoulder, lower arm, upper arm, wrist, and end effector, which was created by a Linear Analysis application. It was noticed that all resulted TFs form has estimated in the fourth order system.

3.2 Response time without controller

The response time curves of the uncontrolled system are presented in **Figure 9**. As shown in **Table 1**, it was noticed that the response time parameters measured in the second unit, that mean the system is very poor responses.

3.3 Optimization proportional-based HSPID

Figure 10 shows the best objectives achieved by IGA. These convergence results reflected positively to optimize PID gains significantly as illustrated in **Table 2**, to be applied on each HSPID controller individually for each manipulator's joint.

3.4 Responses based on the proposed controller

Figure 11 shows the response time results based on each common component HSPID controller with a step signal. The results show a series of system responses to the robot components' orientation angle at a set point. The HSPID controller is a clear reduction response time, but there is some overshoot as illustrated in **Table 3**. The tuned linear responses look satisfactory regarding PID gains based on IGA.

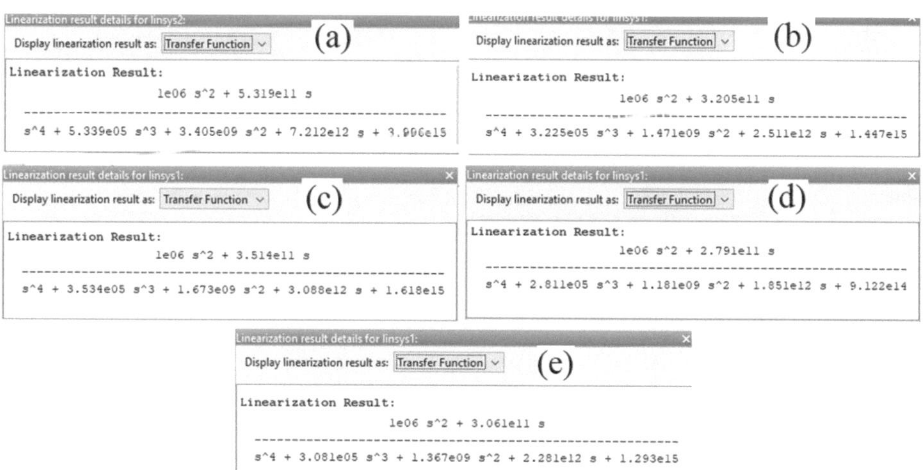

Figure 8.
Resulted estimation TFs of each revolution joint compnents, (a) shoulder, (b) lower arm, (c) upper arm, (d) wrist, (e) end effector.

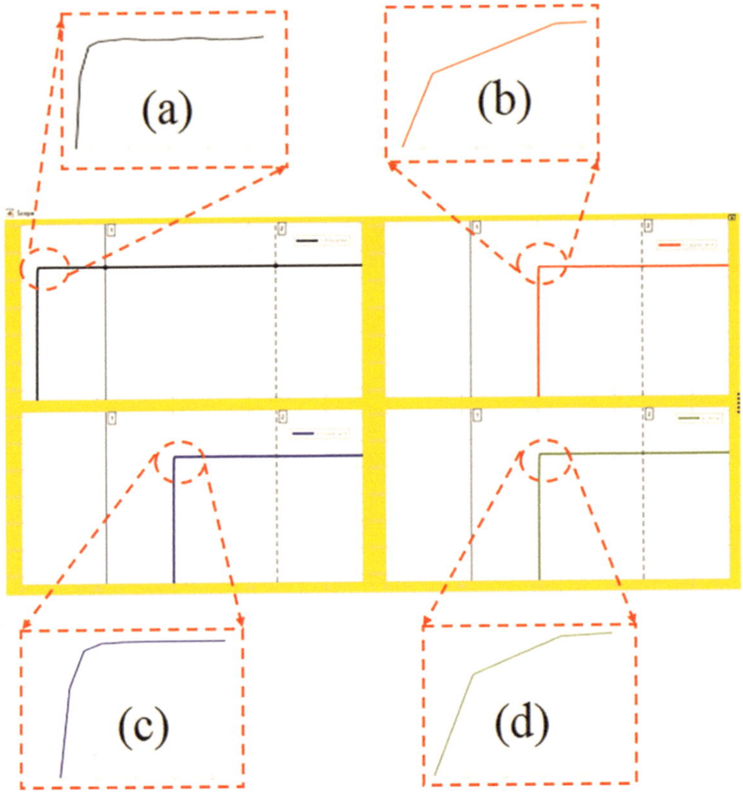

Figure 9.
Response time without controller for each joint components, (a) shoulder, (b) lower arm, (c) upper arm, (d) wrist.

Manipulator Joints	tr(s)	ts(s)	Amplitude
Lower arm	1	1	0.01
Shoulder	0.1	0.1	0.01
Upper arm	1	1	0.01
Wrist	1	1	0.01

Table 1.
Uncontrolled system responses for each joint component.

3.5 Trajectory planning results

As shown in **Figure 12**, three different trajectory signals were injected into the five joint angles to investigate motion case study in x-y plane. The input values to the joints are assorted and the angle motion results were obtained. With the orientation and position vectors as input, the joint angles are obtained as output. A simulated position of the end effector is introduced to be measured the maximum torque for each case on scope simulator. The recorded cases 1,2,3 are 0.06,0.038,0.042 N.m, respectively, as shown in **Figure 13**. Based on the results, it is proven that changing

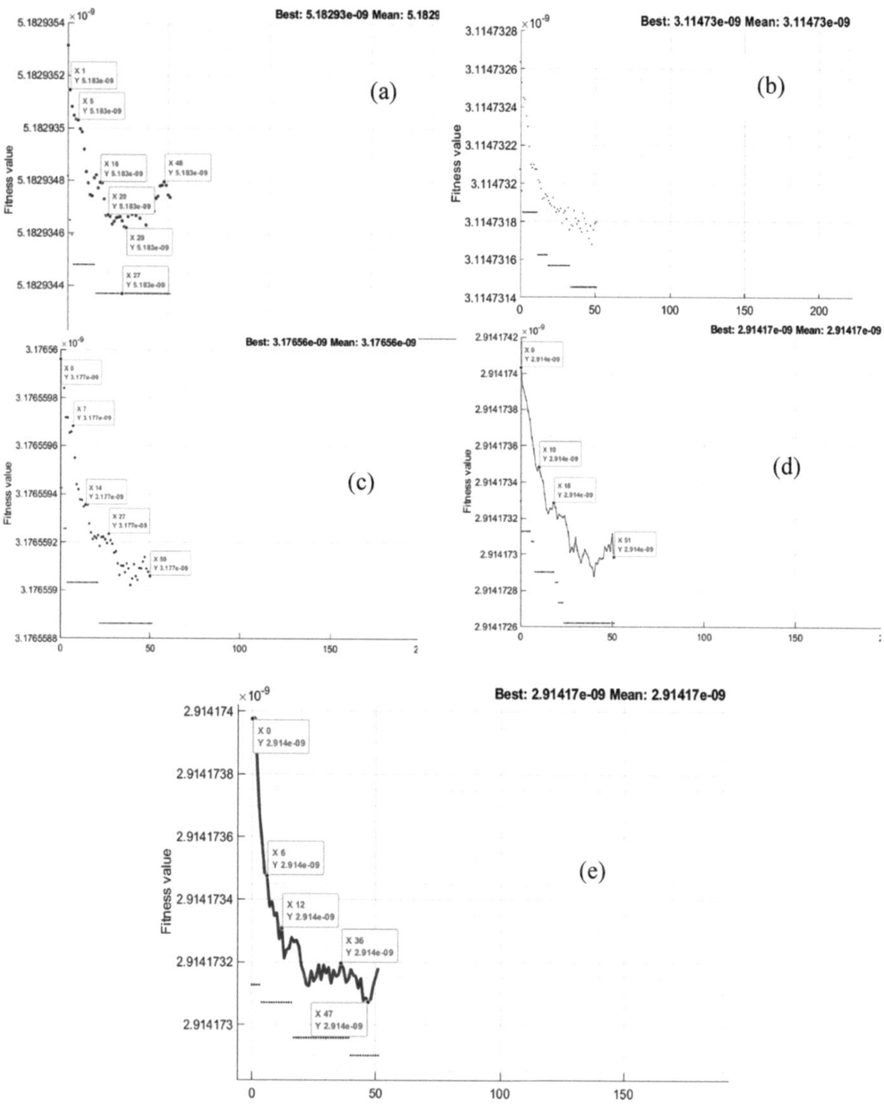

Figure 10.
The effectivness of the IGA on FF for each revolution joint, (a) shoulder, (b) lower arm, (c) upper arm, (d) wrist, (e) end effector.

Manipulator joint	Kp(E8)	Ki (E11)	Kd (E4)	N (E6)	FF(E9)
Shoulder	3.3	7.3	3.3	3.9	5.1
Lower arm	2.4	4.5	3.2	3.8	3.1
Upper arm	3.1	5.9	4	4.7	3.1
Wrist	1.7	2.6	2.6	3	2.9
End effector	4.1	9.4	3.2	3.9	2.9

Table 2.
The optimized proportional gains and FF-based IGA for each revolution joint.

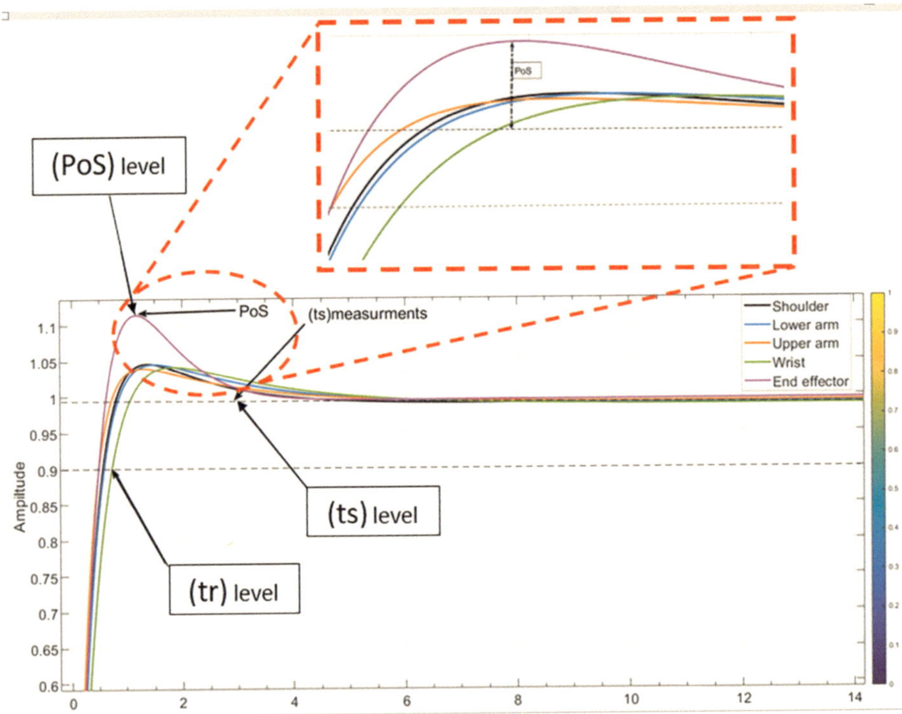

Figure 11.
Case 1 response time reduction based HSPID controller for each revolution joint component.

Joint part	tr (μs)	ts (μs)	Pos
Shoulder	56.2	1352	1.047
Lower arm	58.1	1422	1.044
Upper arm	48.5	1221	1.040
Wrist	73	1130	1.041
End effector	48.1	965.7	1.114

Table 3.
Case 1 response time reduction based HSPID controller for each revolution joint.

the angle of any joint would result in a different end-effector position, and the 3D animation confirms that the arm moves quickly and precisely to the desired configuration.

4. Discussion

To demonstrate the validity of the HSPID controller, we compare the response time characteristics of controlled and uncontrolled systems. The first response comparison includes a robot arm model without a controller, and the second one includes the model-based proposed HSPID controller simulation results **illustrate** that the use

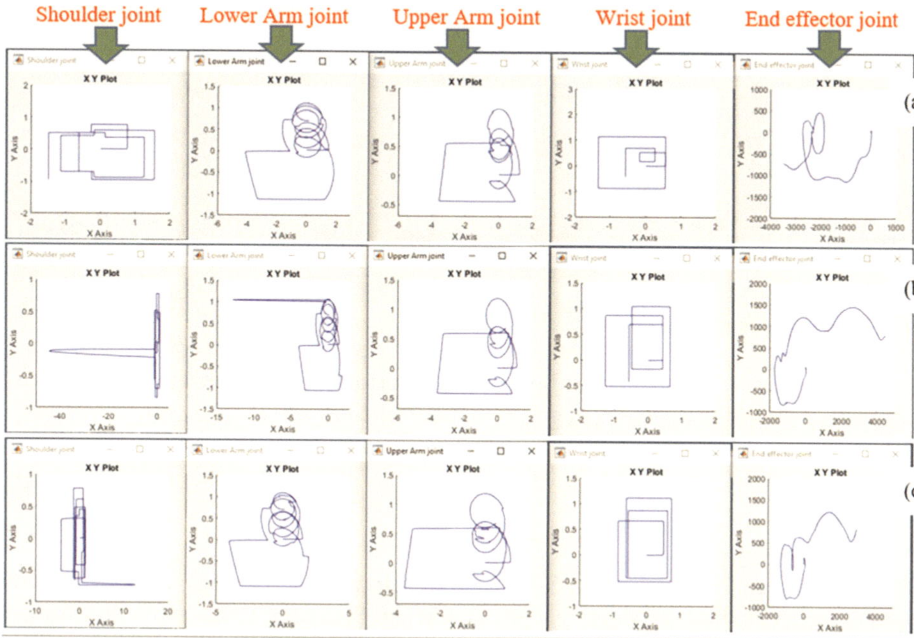

Figure 12.
Trajectory motions of each joint in X-Y plane for three cases, (a) case 1, (b) case 2, (c) case 3.

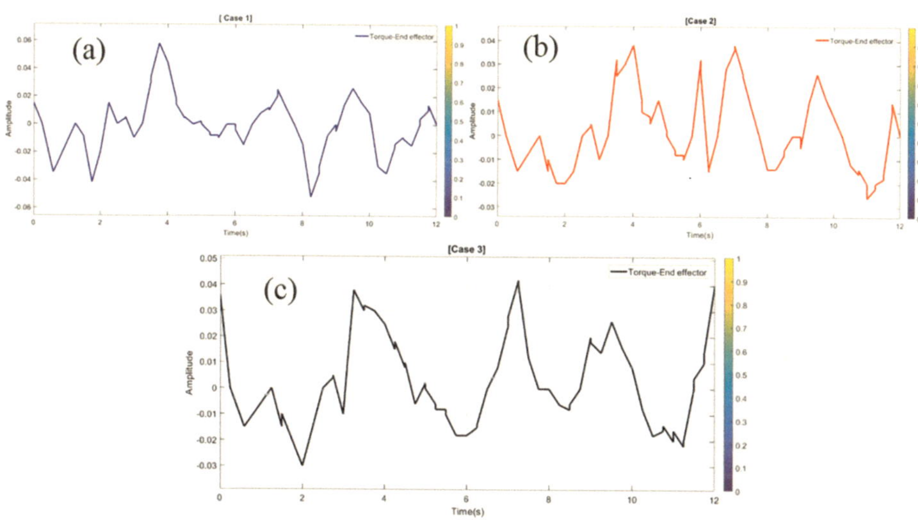

Figure 13.
The torque results of the end effector joint for three case, (a) case 1, (b) case 2, (c) case 3.

of HSPID provides significant reduction. **Table 4** refers to the improved reduction response time ratio (IRRTR) for the tr and ts of the joint manipulators. It can be seen that the maximum IRRTR for the tr occurs at the upper arm and the lowest at the shoulder. The ts results show that the greatest IRRTR occurs at the wrist and lowest at the shoulder. For comparison, the controller can efficiently compensate for orientation errors and quickly settle to an acceptable target value, providing advantages to

Joint components	IRRTR	
	tr (%)	*ts (%)*
Lower arm	17,280	755
Upper arm	20,701	879
Wrist	13,753	950
Shoulder	1850	121

Table 4.
Improved reduction response time ratio (IRRTR).

augmenting the precision of the robot arm. In contrast, researchers can manipulate other parameters in the system to get more analytical results.

5. Conclusion

In this chapter, the innovation 5 DOF robot arm has been designed by the application of SolidWorks side by side with and MATLAB Simscape toolbox for motion analysis and measuring the dynamics of the proposed model. The organigram of the design procedure is divided into five steps:(1) Mechanical configuration and assembly components, (2) Modeling Simscape, (3) Estimation TFs of the manipulator's joints, (4) Designing HSPID controller for joint manipulators (5), Mimic trajectory planning, then analyze response time characteristics and motion result. It is proposed a novel HSPID controller based on IGA offers a best solution to optimize trajectory planning and significant effectiveness of joint's torque.

The motion of the joint angle combinations through various angles and coordinates is controlled by applying three trajectory signals to manipulate the whole system and to compare the paths-based proposed controller with uncontrolled in terms of response time characteristics. For the 12 s trial-tested period, three case studies are and measure the distance, investigate the dynamic motion simulations, and confirm the efficiency of the design. It is observed that the HSPID enhanced the IRRTR significantly for the shoulder, lower arm, upper arm, and wrist in terms of tr by 1850%, 17,280%, 20,701%, and 13,753% respectively, and ts by121%, 755%, 879%, 950%. Thus, the efficiency of the robot arm is confirmed by the case studies, which is relating to trajectory planning. It is observed that the robot arm utilizes torque more effectively. Remarkably, the tr simulation results show that the lowest IRRTR appears into shoulder and the highest into upper arm, where the simulation ts results illustrate that the maximum IRRTR obtains in the wrist and the lowest in the shoulder. The main added value of the study is the elaboration and implementation of the new design method, which offers flexible design and higher-speed motion with less consuming energy. Finally, future manufacturing arm could be greatly improved by applying the proposed innovative design offering the best solution to increase accuracy, speed performance, and boost productivity for wide range of a various tasks.

Conflict of interest

The authors declare that there is no conflict of interest.

Author details

Falih Salih Mahdi Alkhafaji
Ministry of Industry and Minerals, State Company for Electrical and Electronic Industry, Iraq

*Address all correspondence to: falih_alkafaji1@yahoo.com

References

[1] Raza K, Khan TA, Abbas N. Kinematic analysis and geometrical improvement of an industrial robotic arm. Journal of King Saudi University. 2018;**30**(3):218-223

[2] Javaid M, Haleem A, Singh RP, Suman R. Substantial capabilities of robotics in enhancing industry 4.0 implementation. Cognitive Robotics. 2021;**1**(May):58-75

[3] Urrea C, Jara D. Design, analysis, and comparison of control strategies for an industrial robotic arm driven by a multi-level inverter. Symmetry (Basel). 2021;**13**(1):1-20

[4] Zhang B, Liu P. Control and benchmarking of a 7-DOF robotic arm using Gazebo and ROS. PeerJ Computing Science. 2021;7:1-22

[5] A. T, S. M, and V. B. Kinematic analysis of 6 DOF articulated robotic arm. International Research Journal of Multidisciplinary Technovation. 2021; **2027**(1):1-5

[6] Elfasakhany A, Yanez E, Baylon K, Salgado R. Design and development of a competitive low-cost robot arm with four degrees of freedom. Modes in Mechanical Engineering. 2011;**01**(02): 47-55

[7] Clothier KE, Shang Y. A geometric approach for robotic arm kinematics with hardware design, electrical design, and implementation. Journal of Robotics. 2010;**2010**:1-10

[8] Ishak AJ, Soh AC, Ashaari MA. Position control of arm mechanism using pid controller. Journal of Theoretical and Applied Information Technology. 2013; **47**(2):798-806

[9] Rajesh T, Reddy MK, Begum A, Venkatesh D. Design and implementation of robot arm control based on Matlab with Arduino Interface. International Journal of Engineering Research. 2018;**4**(2):43-48

[10] Sánchez AC, Figueroa-rodríguez JF, Fuentes-covarrubias AG, Fuentes-covarrubias R, Gadi SK. Recycling and Updating an Educational Robot Manipulator with Open-Hardware-Architecture Sensors. 2020. pp. 1-22

[11] Myint KM, Min Z, Htun M, Tun HM. Position control method for pick and place robot arm for object sorting system. International Journal of Scientific and Technology Research. 2016;**5**(06):57-61

[12] Ferdinandlivi P, Udaiayakumar KC. Design, modelling and analysis of industrial robotic arm minimizing weight and capital invested. Journal of Critical Review. 2020;**7**(14):639-644

[13] Ansari MJ, Amir A, Hoque MA. Microcontroller based robotic arm: Operational to gesture and automated mode. In: 1st International Conference on Electrical Engineering and Information and Communication Technology, ICEEICT 2014. 2014

[14] Mouli CC. Design and implementation of robot arm control using LabVIEW and ARM controller. IOSR Journal of Electric and Electronic Engineering. 2013;**6**(5):80-84

[15] Aparnathi R, Dwivedi VV. The novel of six axes robotic arm for industrial applications. IAES International Journal of Robotic Automation. 2014;**3**(3): 161-167

[16] Yudha HM, Dewi T, Risma P, Oktarina Y. Arm robot manipulator design and control for trajectory tracking; a review. Proceedings of Electrical Engineering and Computer Science Informatics. 2018;**5**(1):1–6

[17] Rana T, Roy A. Design and construction of a robotic arm for industrial automation. Internaational Journal of Engineering and Research Technology. 2017;**6**(05):919-922

[18] Agrawal R, Kabiraj K, Singh R. Modeling a controller for an articulated robotic arm. Intelligent Control and Automation. 2012;**03**(03):207-210

[19] Gautam R, Gedam A, Zade A, Mahawadiwar A. Review on development of industrial robotic arm. International Research Journal of Engineering and Technology. 2017;**4**(3):1752-1755

[20] Sulaiman I, Tanimu Y. Development of a robot arm: A review development of a robot arm. In: Fed. Polytech. Bida, Sch. Eng. Technol. 8th National Engineering Conference, 2019. pp. 1-5

[21] Alkhafaji FS, Hasan WZW, Sulaiman N, Maryam MBT. Design and implementation a novel system for estimation precise transfer function of DC motor. Advanced Science Technology and Engineering System. 2020;**5**(5):1118-1125

[22] Iqbal J. Modern control Laws for an articulated robotic arm: Modeling and simulation. Engineering Technology and Applied Science Research. 2019;**9**(2): 4057-4061

[23] Alkhafaji F. Modeling and control high speed robotic arm for industrial applications. In: Global Congress on Electrical Engineering (GC-ElecEng2021) 10–12 December 2021At. Valencia, Spain; 2021. pp. 1-10

[24] Kumar Suman S, Kumar Giri V. Genetic algorithms techniques based optimal PID tuning for speed control of DC motor. American Journal of Engineering and Technology Management. 2016;**1**(4):59-64

[25] Alkhafaji FSM, Wan Hasan WZ, Isa MM, Sulaiman N. A response time reduction for DC motor controller using SISO technique. Indonesian Journal of Electrical Engineering and Computer Science. 2020;**17**(2):895-906

[26] Alkhafaji FSM, Hasan WZW, Isa MM, Sulaiman N. Robotic controller: ASIC versus FPGA - a review. Journal of Computational and Theoretical Nanoscience. 2018;**15**(1):1-25

[27] Alkhafaji FSM, Hasan WZW, Isa MM, Sulaiman N. Proposed a novel method for optimization DC motor controller. In: Proc. of the 5th IEEE International Conference on Smart Instrumentation, Measurement and Applications (ICSIMA). 2018. pp. 28-30

[28] Shamshirgaran A, Javidi H, Simon D. Evolutionary algorithms for multi-objective optimization of drone controller parameters. In: CCTA 2021-5th IEEE Conference on Control Technology and Applications. 2021. pp. 1049-1055

[29] Alkhafaji FSM, Hasan WZW. A novel method for tuning PID controller. Journal of Telecommunication Electronic Computer Engineering. 2018; **10**(1–12):33-38

[30] Zhao ZQ, Liu SJ, Pan JS. A PID parameter tuning method based on the improved QUATRE algorithm. Algorithms. 2021;**14**(6):1-14

[31] Ayten KK, Dumlu A. Implementation of a PID type sliding-mode controller design based on

fractional order Calculus for industrial process system. Elektron. ir Elektrotechnika. 2021;**27**(6):4-10

[32] Mpanza LJ, Pedro JO. Optimised tuning of a pid-based flight controller for a medium-scale rotorcraft. Algorithms. 2021;**14**(6):1-24

[33] Mahmood Al-Rawi OY. Enhancing control systems response using genetic PID controllers. Genetic Algorithms in Applications. 2012:35-58

[34] Alkhafaji FSM, Hasan WZW, Sulaiman N, Isa MM. A novel PID robotic for speed controller using optimization based tune technique. Computational Optimization Techniques and Applications Employed. 2021;**32**: 1-22

[35] Wang Z, Zhang Y, Yu P, Cao N, Dintera H. Speed control of motor based on improved glowworm swarm optimization. Computer Material Continuation. 2021;**69**(1): 503-519

[36] Mahfoud S, Derouich A, Ouanjli NEL, Mahfoud MEL, Taoussi M. A new strategy-based pid controller optimized by genetic algorithm for dtc of the doubly fed induction motor. MDPI - Systems. 2021;**9**(2):1-18

[37] G. M. Design and optimization of PID controller using genetic algorithm. International Journal of Research Engineering and Technology. 2015; **02**(06):926-930

[38] Gabis AB, Meraihi Y, Mirjalili S, Ramdane-Cherif A. A Comprehensive Survey of Sine Cosine Algorithm: Variants and Applications. Netherlands: Springer; 2021

[39] Sheta A, Braik M, Maddi DR, Mahdy A, Aljahdali S, Turabieh H. Optimization of PID controller to stabilize quadcopter movements using Meta-heuristic search algorithms. Applied Sciences. 2021;**11**(14)

[40] Aydogdu O, Akkaya R. Design of a Real Coded GA Based Fuzzy Controller for Speed Control of a Brushless DC Motor. London, UK: IntechOpen; 2016. pp. 63-84

[41] Kamal MM, Mathew L, Chatterji S. Speed control of brushless DC motor using intelligent controllers. In: Inspiring Engineering and Systems for Global Sustainability, SCES. 2014

[42] Car J. An introduction to genetic algorithms. Artificial Life. 2014;**3**(1):63-65

[43] Harish Kiran S, Subramani C, Dash SS, Arunbhaskar M, Jagadeeshkumar M. Particle swarm optimization algorithm to find the location of facts controllers for a transmission line. In: Proceedings of International Conference on Process Automation Control and Computing, PACC. 2011. pp. 1–5

[44] Korkmaz M, Aydoğdu Ö, Doğan H. Design and performance comparison of variable parameter nonlinear PID controller and genetic algorithm based PID controller. In: International Symposium on Inovations in Intelligent SysTems and Applications, INISTA. 2012. pp. 1–5

[45] Aly A. PID parameters optimization using genetic algorithm technique for electrohydraulic servo control system. Intelligent Control and Automation. 2011;**02**(02):69-76

[46] Jayachitra A, Vinodha R. Genetic algorithm based PID controller tuning approach for continuous stirred tank reactor. Advanced Artificial Intelligence. 2014;**2014**:1-8

[47] Suresh P, Aspalli MS. Genetic tuned PID controller based speed control of DC motor drive. International Journal of Engineering Trends and Technology (IJETT). 2014; **17**(2):88-93

[48] Arora S, Singh S. Butterfly optimization algorithm: A novel approach for global optimization. Soft Computing. 2019;**23**(3):715-734

[49] Mohamed AAS, Berzoy A, Mohammed O. Control parameters optimization for PM DC motor in photovoltaic applications. In: IEEE International Electric Machines and Drives Conference, IEMDC. 2015. pp. 1742-1747

[50] Hassanat AB, Prasath VBS, Abbadi MA, Abu-Qdari SA, Faris H. An improved genetic algorithm with a new initialization mechanism based on regression techniques. Infection. 2018;**9**(7)

[51] Aguila-Leon J, Chiñas-Palacios C, Vargas-Salgado C, Hurtado-Perez E, Garcia EXM. Particle swarm optimization, genetic algorithm and grey wolf optimizer algorithms performance comparative for a DC-DC boost converter PID controller. Advanced Science Technology Engineering System. 2021;**6**(1):619-625

[52] Chen GY, Perng JW. PI speed controller design based on GA with time delay for BLDC motor using DSP. In: 2017 IEEE International Conference on Mechatronics and Automation, ICMA. 2017. pp. 1174-1179

[53] Kumari S, Prince P, Verma VK, Appasani B, Ranjan RK. GA based Design of Current Conveyor PLD controller for the speed control of BLDC motor. In: Computational Intntelligence and Communication Technology, CICT. 2018. pp. 1-3

[54] Ahmmed T, Akhter I, Rezaul Karim SM, Sabbir Ahamed FA. Genetic algorithm based PID parameter optimization. American Journal of Intellectual System. 2020; **10**(1):8-13

[55] So GB. A modified 2-DOF control framework and GA based intelligent tuning of PID controllers. PRO. 2021; **9**(3):1-19

[56] Apriaskar E et al. Microwave heating control system using genetic algorithm-based PID controller. IOP Conference Series in Earth Environmental Science. 2022;**969**(1):1-10

[57] Nasri M, Nezamabadi-Pour H, Maghfoori M. A PSO-based optimum design of PID controller for a linear brushless DC motor. World Academy of Science, Engineering and Technology. 2007;**26**(40):211-215

[58] Bhatt K, Bundele M. Review paper on PSO in workflow scheduling and cloud model enhancing search mechanism in cloud computing. IJIET-International Journal of Innovation. 2013;**2**(3):68-74

[59] Nabab M. Particle swarm optimization: Algorithm and its codes in MATLAB. ResearchGate. 2016;**1**:8-12

[60] Freitas D, Lopes LG, Morgado-Dias F. Particle swarm optimisation: A historical review up to the current developments Diogo. Entropy. 2020; **22**(3):1-36

[61] Mohd Zakki MI, Mohd Hussain MN, Seroji N. Implementation of particle swarm optimization for tuning of PID controller in Arduino Nano for solar MPPT system. International Journal of Electrical and Electronic System Research. 2018;**13**(11):1-8

[62] Bagyaveereswaran V. Particle swarm optim controller for Mppt. 2018;**9**(12): 1057-1065

[63] Xie Y, Meng J. PID control for the vehicle suspension optimized by the PSO algorithm. 2018;**2018**:172-177

[64] Arain BA, Shaikh MF, Harijan BL, Memon TD, Kalwar IH. Design of PID controller based on PSO algorithm and its FPGA synthesization. International Journal of Engineering and Advanced Technology. 2018;**8**(2):201-206

[65] Howimanporn S, Chookaew S, Sootkaneung W. Implementation of PSO Based Gain-Scheduled PID and LQR for DC Motor Control Using PLC and SCADA. In: 2018 International Conference on Control and Robots, ICCR. 2018. pp. 52-56

[66] Nazelan AM, Osman MK, Samat AAA, Salim NA. PSO-based PI controller for speed Sensorless control of PMSM. Journal of Physics Conference Series. 2018;**1019**(1)

[67] Latha K, Rajinikanth V, Surekha PM. PSO-based PID controller Design for a Class of stable and unstable systems. ISRN Artificial Intelligence. 2013;**2013**:1-11

[68] Singh R, Kuchhal P, Choudhury S, Gehlot A. Design and experimental evaluation of PSO and PID controller based wireless room heating system. International Journal of Computers and Applications. 2014;**107**(5):15-22

[69] Xiang Z, Ji D, Zhang H, Wu H, Li Y A simple PID-based strategy for particle swarm optimization algorithm. Information Science. 2019;**502**:558-574

[70] Hassan R, Cohanim B, De Weck O, Venter G. A comparison of particle swarm optimization and the genetic algorithm. In: 46th AIAA/ASME/ASCE/AHS/ASC Structures, Structural Dynamics & Materials Conference. 2005. pp. 1138-1150

[71] Bensalem S, Ingrand F, Sifakis J. Autonomous robot software design challenge. In: Sixth IARP-IEEE/RAS-EURON Joint Workshop on Technical Challenge for Dependable Robots in Human Environments. 2008. pp. 1–5

[72] Shukla RK, Deshmukh DB. A review on role of CAD / CAM in designing for skill development. International Journal of Research and Engineering Science Technology. 2015;**1**(June):2016

[73] Aburaia M, Markl E, Stuja K. New concept for design and control of 4 axis robot using the additive manufacturing technology. Procedia Engineering. 2015; **100**(January):1364-1369

[74] Rodríguez E et al. Analysis of Robotic System Motion in SimMechanics and MATLAB GUI Environment. London, UK: IntechOpen; 2014. pp. 565-581

[75] Jadeja Y, Pandya B. Design and development of 5-DOF robotic arm manipulators. 2019;**8**(11):2158-2167

[76] Pawar V, Bire S, More S. Review on design and development of intelligent robotic arm Generation-1. International Journal of Innovation Science and Research Technology. 2018; **3**(3):527-529

[77] Ebrahimi N. “Modeling, Simulation and Control of a Robotic Arm.” 2019. pp. 1–7

[78] Sabri M, Fauzi R, Fajar MS, Geubrina HS, Sabri FAM. Model and simulation of arm robot with 5 degrees of freedom using MATLAB. IOP Conference Series Materials Science and Engineering. 2021;**1122**(1):012032

[79] Gasparetto A, Seriani S, Scalera L. Modelling and control of mechatronic and robotic systems. Applied Sciences. 2021;**11**:4

[80] Llopis-Albert C, Rubio F, Valero F. Modelling an industrial robot and its impact on productivity. Mathematics. 2021;**9**(7)

[81] Alkhafaji FSM, Hasan WZW, Isa MM, Sulaiman N. A Modified GA based PI controller for DC Motor Performance. In: Proc. of the 6th IEEE International Conference on Smart Instrumentation, Measurement and Applications (ICSIMA). 2019. pp. 1–4

[82] Benotsmane R, Dudás L, Kovács G. Trajectory optimization of industrial robot arms using a newly elaborated 'whip-lashing' method. Applied Sciences. 2020;**10**(23):1-18

[83] ALkhafaji FSM, Hasan WZW, Isa M, Sulaiman N. Prime Asia2019. In: A HSMDAQ System for EstimatingTransfer Function of a DC motor, Prime Asia. 2019. pp. 25-28

[84] Urrea C, Cortés J, Pascal J. Design, construction and control of a scara manipulator with 6 degrees of freedom. Journal of Applied Research Technology. 2016;**2**:396-404

[85] Amr Nasr A, Gaber E, Rezeka SF. Design and position control of arm manipulator; experimentally and in MATLAB Sim mechanics. International Journal of Engineering Research and Technology. 2016;**5**(8):352-359

[86] Carpio M, Saltaren R, Viola J, Calderon C, Guerra J. Proposal of a decoupled structure of fuzzy-pid controllers applied to the position control in a planar cdpr. Electronics. 2021;**10**(6):1-21

[87] Yura J, Oyun-Erdene M, Byambasuren BE, Kim D. Modeling of violin playing robot arm with MATLAB/ SIMULINK. Advanced Intellectual System Computing. 2017;**447**(January): 249-261

Chapter 8

Healthcare Robots and Smart Hospital Based on Human-Robot Interaction

Kazuhiko Terashima, Kazuhiro Funato and Takuyuki Komoda

Abstract

This chapter first introduces the research on next-generation care systems and stations that the author's group has actually conducted. Next, the development trends, challenges, and prospects for smart hospitals, which aim to improve overall hospital efficiency based on CPS technologies such as sensing, IoT, AI, and big data processing, in addition to robotics and control technologies, toward the realization of Society 5.0, are described. Third, the concept of the smart hospital and discussion will be explained to provide basic knowledge for its construction. Finally, future scope and conclusion will be described.

Keywords: smart hospital, health care system/station, human-robot symbiosis, health care robot, social robot

1. Introduction

Society 5.0 is a society in which humans can lead more comfortable and vibrant lives with the help of AI and robots [1]. This new type of society proposed in the Fifth Science and Technology Basic Plan in January 2016 includes the Internet of Things (IoT) [2], big data [3], artificial intelligence (AI), robots, and other cyber-physical systems (CPS) [4, 5]. The basic technologies are listed below. Along with the smart society and the smart factory, the smart hospital is one of the targets. Having established the Human-Robot Symbiosis Research Center at Toyohashi University of Technology in 2010, the lead author has been engaged in robotics research related to human-robot symbiosis, as well as research on next-generation nursing care systems and stations as a project of Japan's Ministry of Education, Culture, Sports, Science and Technology. The following robots are considered to perform major work processes for reducing care work and ensuring greater patient comfort: a patrol robot, an integrated transfer/transportation robot, an omnidirectional gait support robot, a power-assisted bed, a social media robot, and a comprehensive system for care scheduling and care. We have been conducting research and development of the station etc. Therefore, this chapter first introduces the research on next-generation nursing care systems and stations that the author's group has actually conducted. Next, the authors discuss the trends, challenges, and prospects for the development of smart hospitals that are intended to optimize the efficiency of entire hospitals based on CPS technologies such as sensing, IoT, AI, and big data processing, in addition to robotics and control technologies, with the aim of realizing Society 5.0. The concept of a smart hospital is described, providing basic knowledge for the construction of smart hospitals.

2. Research trends and results of next-generation care station development

2.1 What is the next-generation care station

Figure 1 shows the main tasks in which robots could be used in nursing homes and hospitals. There are various types of tasks involved in assisting caregivers. However, it would be wasteful to introduce robots only for specific tasks if sufficient use were not made for those robots. It is necessary to clarify the position of robots in nursing care and daily life support, and to introduce robots and systems as a comprehensive system, taking into consideration the coordination of each task. In addition, showcasing of care systems and care stations that demonstrate the workflow of human-robot symbiosis and coexistence would be beneficial for future practical use. Although research on smart spaces and robot houses [6] has been undertaken in recent years, there have been few showcases for nursing care. They include the Ochanomizu University Ubiquitous Computing Experimental House (aka Ochahouse) [7], where sensors are embedded in the space and the residence is fully equipped with a sensing system, and the Human Residence House (C-PRH), where research is being conducted on optimization of the residential environment for the introduction of robots. In addition, SELF (Sensorized Environment for LiFe) [8], and robotic rooms are being studied, which are useful for the automation of nursing care [9]. However, unlike in an automobile factory, these systems are not intended to be completely unmanned. Rather, the goal is to create a comfortable and safe system conducive to cooperation and coexistence among medical personnel, patients, and robots. The

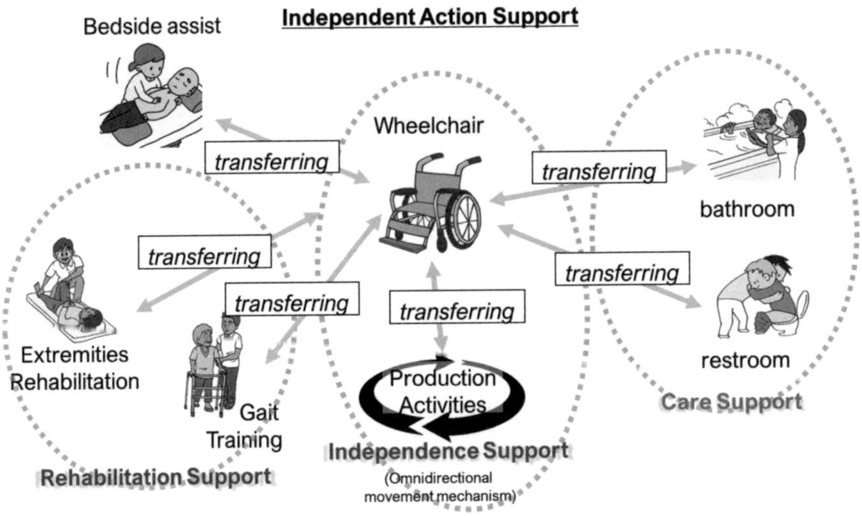

Figure 1.
Examples of major tasks for which robots may be used in nursing homes and hospitals. For example, support for independence, including rehabilitation and daily life support from the bedside.

Research Center for Human-Robot Symbiosis at Toyohashi University of Technology has been working on the project "Development of Next-Generation Nursing Care System and Station to Enrich the Super-aging Society," which was specially funded by the Ministry of Education, Culture, Sports, Science and Technology (FY 2012–2018). We have been conducting innovative and unique research and development with the aim of developing a system enabling the realization of high-quality services that provide a sense of healing and hospitality so that people can experience a sense of purpose in their lives. Our group's robot is distinctive in that it is not a fully automatic robot, but an assistive robot focused on human intention that is easy to operate and use virtually anywhere, giving a sense of security, and aiming for human-robot symbiosis. Therefore, we first introduce some representative care robots and care systems that have been researched and developed by the authors' group [10].

2.2 Hospital rounds robot Terapio and Patrol robot Kurumi

We are developing a mobile medical checkup robot that performs the two tasks of moving *things* and *information*. We have developed Terapio, a medical rounds robot that replaces the conventional rounds carts used by doctors and nurses during rounds in hospitals, and supports the transportation of medical tools and examination work (recording of medical data) (**Figure 2**) [11–13]. As a result of examining how to operate the robot as a medical robot, the need for omnidirectional movement was considered, especially for bedside maneuvering. Mechanisms capable of omnidirectional movement include the free-roller type, ball type, and caster type, typified by the omniwheel and the mecanum wheel. The free-roller type transmits power to the road surface by means of free rollers on the periphery of the wheel. Because of its simple structure, low cost, and relatively easy control, it is still used as a power source for general omnidirectional moving platforms. However, because of the free rollers, this mechanism has a low ability to go over bumps and traverse uneven terrain. In addition, vibration and noise problems may make it unsuitable for some applications. The DDSS (Differential Drive Steering System) (**Figure 3**) [14–16], a unique

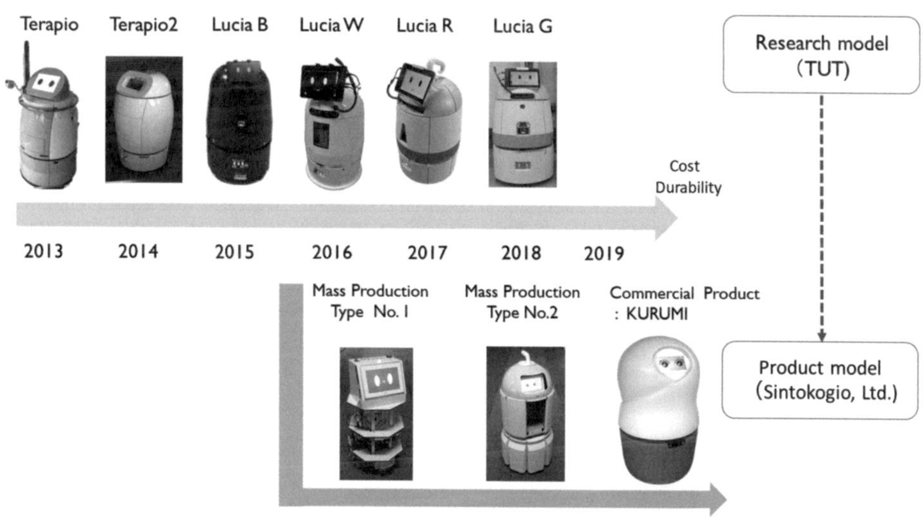

Figure 2.
Transition of mobile robots developed by the authors. The upper row shows basic research including Terapio, a medical round robot. The bottom row shows applied research for practical use including KURUMI.

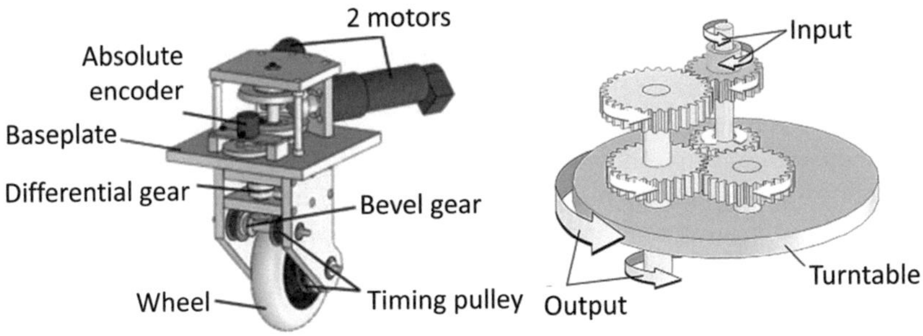

Figure 3.
Structural drawing of the DDSS developed by the authors, an omni-directional wheel that maximizes the efficiency of two motors and can move instantly in any direction.

differential drive steering mechanism developed by the authors, is used as the drive system for this robot and equipped with an omnidirectional movement function that is both quiet and has high performance over bumps. The main body of the robot is equipped with a range sensor that recognizes the environment in all lateral directions as well as forward and downward. The robot uses the range sensor to recognize the environment and has a tracking function to autonomously follow a specific person while avoiding obstacles. This eliminates the need to prepare map information and a travel plan in advance. The robot also has a ring-shaped power-assisted handle that can be operated accurately with a small amount of force when moving Terapio by direct touch to enable other people to position it at the bedside. The robot together with the object to be transported weighs approximately 80 kg. Without power assist, the user is subjected to a large load, but with power assist, the operation is possible with a small load of 10–20 N. In addition, it has a function that allows patients' personal information and vital data to be recorded and viewed, including past information, and at the nurses' station, it has a function that allows sending/receiving rounds data and managing all patient data by a wired connection to an information server. User authentication is provided to prevent unauthorized viewing and manipulation of rounds and patient data. The database is encrypted and cannot be deciphered by direct access. Terapio's overall integrated system can be switched by the supervisory controller to autonomous tracking mode, power-assisted mode, and consultation mode for different types of robot operations. As shown in **Figure 2**, in the Aichi Prefecture "Priority Research Project in Knowledge Hub Aichi" from FY2017 to FY2019, the Toyohashi University of Technology and Sintokogio, Ltd. and nursing home of the Tenryu Koseikai social welfare corporation were used as field test sites for practical use in cooperation with the private sector. The robot Kurumi was developed as a night patrol robot in a nursing home to manage people with dementia. It is based on Terapio's basic technology and is equipped with a function that enables it to move around the facility automatically and to contact caregivers after detecting a wandering person with dementia. The company is planning to commercialize the product.

2.3 Transfer and transportation integrated care robot

Until now, two types of nursing care equipment, a patient lift, and a wheelchair have been required to assist with transfers and transportation in the nursing care field. Therefore, we analyzed these processes and proposed and fabricated an omnidirectional power-assisted transfer/transportation integrated care robot that combines an omnidirectional transfer mechanism and a power-assisted lift mechanism, with the aim of performing these tasks with a single robot (**Figure 4**) [17, 18]. A mechanism that can be transformed to the required size at the time of use was designed to enable use in the home. The minimum height is 1750 mm × 720 mm width × 1050 mm depth, and the maximum height is 2100 mm × 720 mm width × 1600 mm depth. Using the prototype integrated transfer/transportation care robot, a series of care actions were realized, including attaching a sling sheet to a patient in bed, lifting and lowering the patient in the lift section and moving the patient toward the seat, placing the patient in the seat, and transporting the patient in the transfer section of the integrated robot (**Figure 5**). Experiments have demonstrated that a single robot can perform two major tasks and is mobile, making it innovative and convenient to use virtually anywhere. The robot is characterized by the use of multiple power-assist

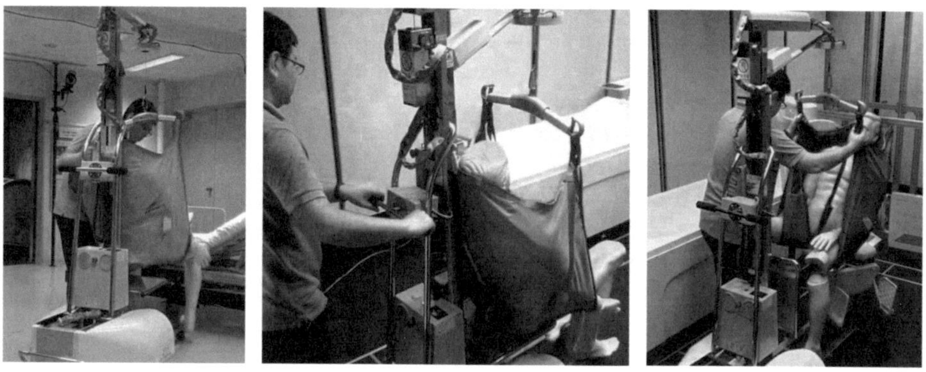

(a) Transferring from bed to wheelchair (b) During Transportation from bed to bath (c)Transfer from wheelchair to bathing tub

Figure 4.
The authors have developed a robot that integrates transport and transfer. Power-assisted operation is used for lifting and lowering and moving the robot so that anyone can use it easily.

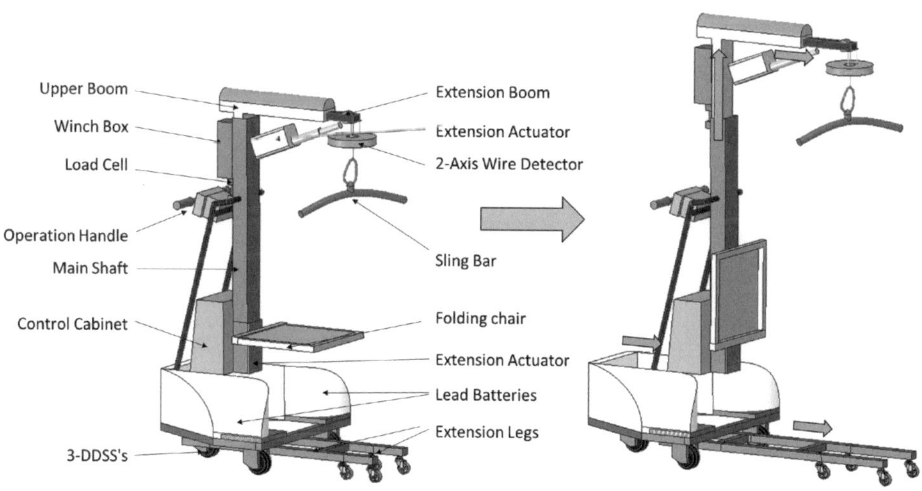

Figure 5.
Image of the deformation mechanism of a robot that integrates transport and transfer. The mechanism expands during transfer operations and contracts when the robot is not being moved or used.

technologies for the desired task. For example, for patient lifting operations, power-assisted movements are realized using multiple sensors built into the wires of the lift. One technology employed is power assist using a tension-type load cell built into the wire traction section. The load on the wire is detected by the load cell, and the speed of the traction motor is controlled based on the displacement to perform lifting. In particular, the controller has been designed to suppress the limit cycle that occurs during descent work when the machine is grounded, thereby achieving smooth control. Another technology employed is an optical sensor that detects the tilt angle of the lift's wires without contact and moves the main body of the robot based on displacement. This can be done seamlessly with the patient lifting operation, allowing the caregiver to move the robot body in any direction and at any speed while supporting the patient with both hands. This allows the caregiver to use both hands to support the patient, which gives the patient a sense of security, unlike pendant-type switches, and because both hands are used for support, there is less swaying and thus safety is enhanced. Furthermore, by employing the aforementioned DDSS in the robot's movement mechanism and mounting a control handle with a force sensor on the backrest, the robot body can move in all directions by power-assist control, enabling operation in the direction of travel and speed as intended by the caregiver. It is not as complicated to control as automatic driving and extremely easy to operate. The results of this integrated care robot research were discussed with medical professionals of Toyohashi Narita Hospital. They commented on the convenience of using a single device to perform two processes. In addition, noting that the power-assist system reduces the workload, they expressed a strong desire to use it. One medical professional commented that he would like to use the robot in the transfer and transportation of patients from the bed to the bathing area.

2.4 Walking training robot combining Niltwamor and Lucia

As a consequence of population aging, the number of elderly people with gait disorders is increasing. Early rehabilitation is important to prevent the elderly from becoming bedridden and to restore their walking ability. The authors have developed two walking robots, Niltwamor (Novel Intelligent Walking Mobile Robot) [19–22] and Lucia [22, 23] (**Figure 6**). Niltwamor consists of an omnidirectional drive system using an omniwheel and a laser sensor, and an unloaded weight support system that suspends the harness independently by two wires. Rehabilitation is considered to be effective when started early after treatment and is particularly suitable in the acute phase when the body is difficult to support. In order to increase the degree of freedom in the direction of gait training, a control system is designed to determine the direction in which the gait training robot moves and the distance it travels by monitoring the direction and step width with a laser sensor when the trainee takes a step. A feature of this system is that the trainee has a high degree of freedom to move in the direction in which he or she wants to go, rather than being restricted to a fixed location as in the case of a conventional treadmill. Since excessive floor reaction force causes recurrence of disability and delay in recovery, and excessive weight bearing decreases the effectiveness of rehabilitation, we proposed an adaptive control system that uses sensors to estimate floor reaction force and control the lifting force to keep the floor reaction force appropriately constant or arbitrary. To evaluate the effectiveness of the proposed gait training robot, Niltwamor was applied to elderly patients undergoing gait rehabilitation. The results showed that the robot operated almost exactly as the patient intended. The physiotherapist commented that the data

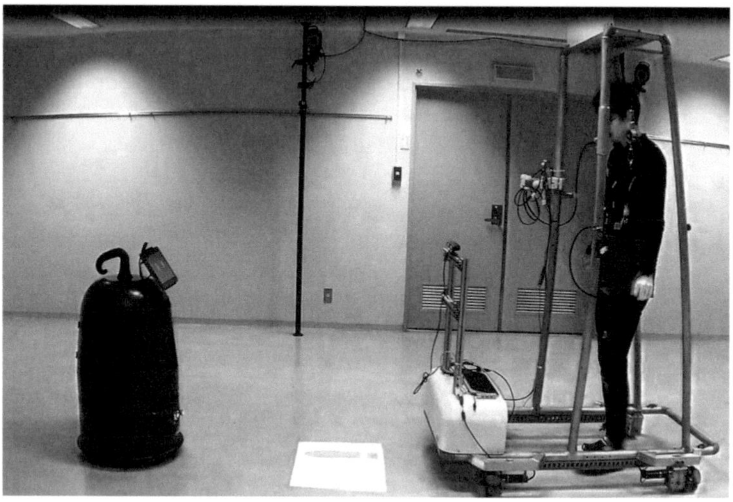

Figure 6.
Two robots developed by the authors cooperate in a gait training task, with Lucia (left) providing gait guidance and Niltwamor (right) performing real-time unloaded weight support system that suspends the harness independently by two wires.

obtained online from the sensor, which measures the unloading force and floor reaction force, enables statistical and theoretical analysis of the relationship with the treatment effect, which was previously done empirically, and is effective for use as digital therapy. Lucia, on the other hand, is considered particularly effective for rehabilitation during the recovery period. The robot's approach to measuring and intervening in whole-body movements is promising for motor function rehabilitation. The Robotics Center at TUT has focused on the importance of supporting motor recognition during the recovery process of motor function and has designed a robot system that enhances motor recognition through interaction between a human and a robot. We constructed a system that measures the gait pattern of a trainee using laser sensors and cameras, evaluates the gait condition using digital data and a skeleton model, and determines the gait condition such as a slip gait, a half-turn gait, and a small step gait. The ideal steps of gait are projected on the floor from Lucia by a laser point using image projection control to navigate the gait training. When the gait deviates from the ideal, the vibrator attached to the waist can be controlled to vibrate and stimulate the senses, such as voice, to encourage the user to walk. Experiments involving patients with Parkinson's disease and cerebral palsy verified the effectiveness of the proposed method by examining the advantages of digital motion measurement and sensory stimulation. In the acute phase, Niltwamor can be used by itself; in the next phase, Lucia can be used together with Niltwamor as a navigational system, as shown in **Figure 6**; and finally, in the recovery phase, Lucia alone can be used for gait training, for example, and these two paired robots can be used to train gait step by step. The system can be used for a wide range of applications.

2.5 Power-assisted bed

When moving a bed from a room to a corridor, it is difficult for a single nurse to move the bed due to the weight and maneuverability of the bed, and turning a bed with ordinary wheels in a small room is a challenging task. A bed-cum-wheelchair called "resyone" has recently been developed, which is unique in that a part of the bed can be removed as a wheelchair for transfer, but the task of detaching the wheelchair from the bed takes a significant amount of time. The author's group has developed a system in which a handle with a force sensor is attached to the rear of the bed, and the direction of the pushing force is used to determine the direction in which the user wants to go, and the speed of bed transportation is determined by the magnitude of the force, taking into account the habits of human operators and automatically controlled by AI [24–26] (**Figure 7**). Omnidirectional wheels allow the machine to move instantly in any direction. This enables a single nurse to transport a patient in the desired direction and at the desired speed without requiring a great deal of force or turning movements.

2.6 The concept of "weak robots" and social robots in a human-robot symbiotic society

Good coordination between humans and robots is necessary for a society where humans and robots coexist, not to mention in nursing care facilities. Professor Michio Okada of Toyohashi University of Technology has proposed the concept of "weak robots" and has developed a theory and applied it in various fields. He believes that if a robot is too close to perfection and too strong, people are less likely to approach it and that a robot with imperfect elements is often more suitable for daily life [27–29]. For example, unlike Roomba, the Trash Can Robot does not do all the work itself. Children who see it tend to become friendly toward it and actively try to help the robot by picking up trash. Such a response is an important aspect of a rich human life. Extending this idea, we have developed a communication robot Moo and other robots with the aim of stimulating communication and collaboration between caregivers and

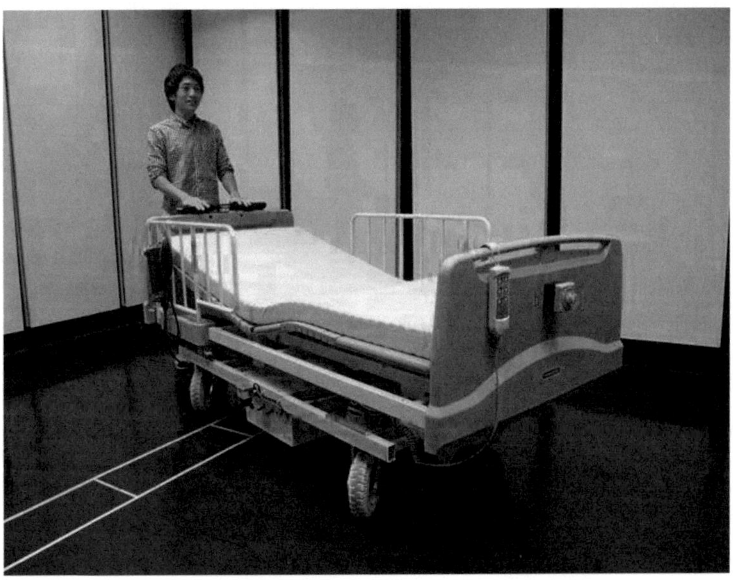

Figure 7.
Power-assisted nursing care bed developed by the Authors can be moved in Omnidirectional. The traveling cart, which is add-on to a commercially available care bed, can be easily moved back and forth, left and right, and swiveled by a single person by lightly applying force to the operating handle.

DOI: http://dx.doi.org/10.5772/intechopen.107070

those who need care. Expanding on these studies, we are currently designing and developing various interfaces for social robots (**Figure 8**).

2.7 Care systems and care stations that have been developed

2.7.1 Care scheduling system that coordinates caregivers and care robots

As an example of a care system, we introduce our approach to care scheduling (**Figure 9**) [30, 31]. For people requiring nursing care to use nursing care facilities safely and comfortably, it is important to be able to provide the services desired by the users (persons requiring nursing care) on a just-in-time basis. On the other hand, as the population of people requiring nursing care increases, the burden on caregivers is increasing year by year. Caregiving is hard work, both physically and mentally, and there is an urgent need to reduce this burden as much as possible. In other words, there is a need to create a care plan that improves patient satisfaction and reduces the burden on caregivers. The authors' group has therefore applied combinatorial auction theory to construct a care planning support system that automatically creates a schedule, takes into account the user's preferences for services and caregivers, and determines the assignment of caregivers, including collaboration with robots that reduces the caregivers' burden. Care planning is an urgent issue for nursing homes and hospitals, and this care planning support system needs to be verified in the field in the future.

2.7.2 Nursing care station

To create a showcase for care stations, it is important to develop fundamental technology covering simulation software and animation, scheduling and optimization,

Figure 8.
Weak robots—research on social implementation of relational robots. A weak robot does not hide its weaknesses and imperfections. Rather, by disclosing its weaknesses in moderation, the robot successfully draws out the "strengths and kindness" of those around it.

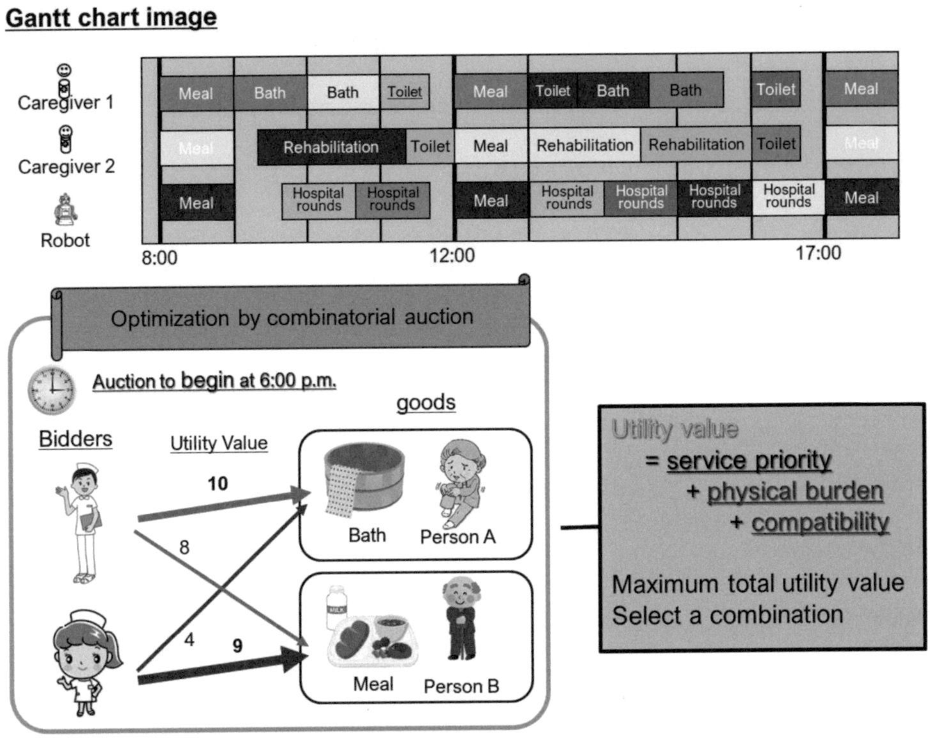

Figure 9.
Care scheduling using combinatorial auctions developed by the Authors. Optimization of resources both improves user satisfaction and reduces the burden on providers.

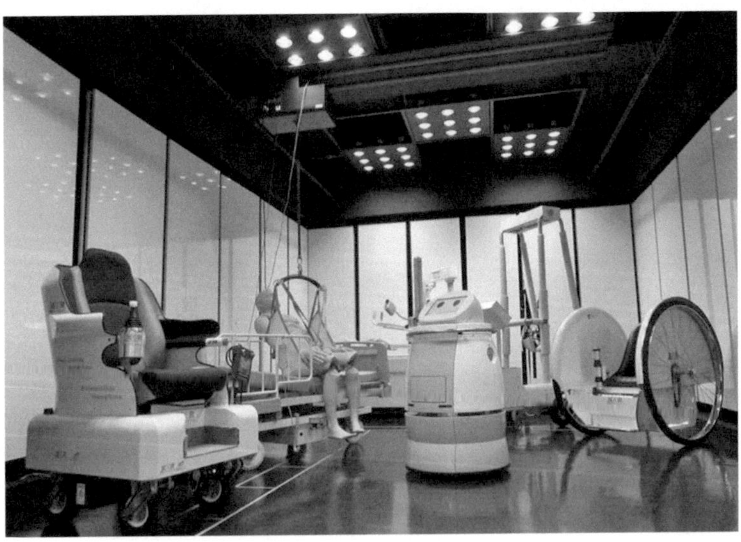

Figure 10.
The authors' prototype nursing station showroom. By placing the various robots developed in the showcase, it is possible to verify not only from the perspective of researchers but also from the perspective of users what kind of robot fits in the nursing care field.

control software for coordinated work of care robots, monitoring and surveillance, and evaluation and redesign. A showcase for care stations is essential for visualizing the concept and presenting the results. This will provide a venue where medical professionals, researchers, engineers, manufacturers, and those involved in sales and marketing can gather at any time and make it easier for them to create usable products. In terms of providing a venue, the development of a showcase for nursing care robots would also be useful. **Figure 10** is a photograph of a showcase that is part of a care station [32, 33]. Our project pursued development of major care robots as a comprehensive system and involved coordination of robots, care scheduling considering the use of robots, assistive robots considering the roles of humans and robots in a human-robot symbiotic society rather than full automation, social media robots, etc. The main elements of our project were included in the presentation of the 2018 Ministry of Education, Culture, Sports, Science and Technology project's final results. We have achieved a certain level of success and will continue development with a view to realization of smart hospitals.

3. Smart hospital initiative

3.1 What is a smart hospital?

A smart hospital [34] is a new type of hospital that combines information technology and medical technology to ensure the quality of medical care in the coming super-aged society, curb the increase in medical costs, improve international competitiveness in the medical field, and reduce the burden on medical personnel. Also called an AI Hospital [35], it is a facility that will serve as a base for medical care aimed at Society 5.0. Progress toward Society 5.0 in the medical field is divided into four categories: Comfortable living, Health promotion, Optimal treatment, and Reduction of burden (**Figure 11**). Medical big data will be created with digitized medical information and patient health information obtained through IoT devices. The aim of this project is to utilize AI technology to improve health management for everyone and treatment outcomes for patients, provide diagnostic assistance to reduce the burden of medical treatment, and support medical education and communication with patients.

Source: Cabinet office: Examples of Creating New Value in the Fields of Healthcare and Caregiving (Society 5.0). https://www8.cao.go.jp/cstp/english/society5_0/medical_e.html

3.2 What the smart hospital aims to achieve

The advantage of smartization is that it enables optimization for individual patients through the use of data. Capitalism up to now has aimed at the happiness of the greatest number of people by raising the average level, but by responding to individual needs through the use of smart technology, it has become possible to achieve the greatest happiness in terms of diversity of outcomes as well as for the greatest number of people. Smart hospitals aim to achieve the four above-mentioned goals, which are readily divisible into two categories. The first concerns the provision of what not only patients visiting the hospital but also everyone in the community wants from the hospital (**Table 1**), and the second concerns reduction of unnecessary work and tasks so that doctors and nurses can better perform their duties (**Table 2**). What everyone wants from hospitals is an integrated experience that makes them feel

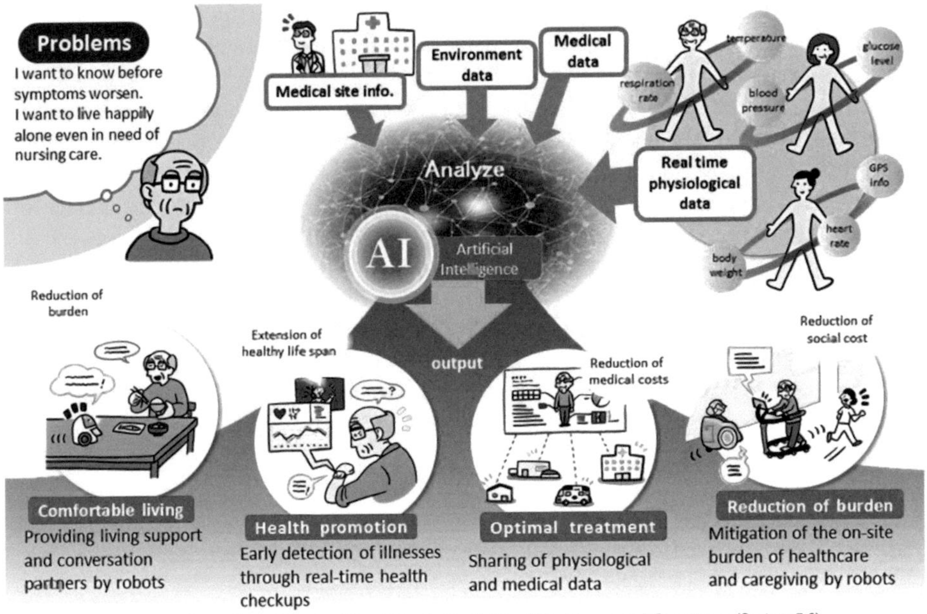

Source: Cabinet office: Examples of Creating New Value in the Fields of Healthcare and Caregiving (Society 5.0)
https://www8.cao.go.jp/cstp/english/society5_0/medical_e.html

Figure 11.
Society 5.0 in the medical field. AI analysis can provide appropriate resources for various requirements such as life support, health promotion, optimal treatment, and burden reduction.

Provide what people need
• Online reception and automated accounting, but convenience alone is not enough
• Provide an integrated experience that makes people want to visit the hospital and be glad they did
• Responsibility as a health information center to manage the health of the community
• Health management from the viewpoint of pre-symptomatic disease through the Internet of Medical Things (IoMT).
• Explanation of treatment details using augmented reality (AR) etc.

Table 1.
What the smart hospital aims to achieve #1.

What helps doctors and nurses perform their duties better
• Online reception, order management service, medical appointment management application
• Speech input, natural language, and AI-based diagnostic assistance
• Patient risk behavior management by IoMT
• Robots to assist in treatment (surgery, catheterization, etc.), patient care and material handling
• Creation of medical education tools using augmented reality (AR) etc.

Table 2.
What the smart hospital aims to achieve #2.

good about visiting a hospital, as well as the ability to manage community health from the early stages of disease, enabling early detection and treatment of illnesses and early reintegration into society. The smooth transmission of various information, such as health management and early detection of disease using a wearable device with electrocardiogram and other functions, medication management using a smartphone application, alert functions such as the date of a return visit and the order of calls, online reception and automatic accounting, etc., will provide convenience for patients and everyone else. All these elements will contribute to the enhancement of the system. In addition, AI interviewing and diagnostic assistance in the field of medical care will not only shorten the time required for medical treatment but also contribute to an accurate diagnosis. Dictation and natural language software can reduce the time spent on manual entry of medical records, allowing doctors and nurses to spend more time face-to-face with patients, which is the essence of medical care; patient behavior monitoring using IoT devices can help manage risky patient behaviors such as accidental falls and self-extubation. The use of robots in healthcare can also reduce the burden of various tasks such as guiding patients and carrying items. It is also effective in providing medical care in the midst of infectious diseases such as the current COVID-19 pandemic and in preventing the spread of such diseases. It has also been reported that in the medium term it will contribute to reducing the cost of healthcare [36]. Thus, the combination of information technology and medical technology is expected to improve the quality of medical care in various situations.

4. Discussion

In the U.S., the introduction of electronic health records (EHRs) progressed owing to the promotion of IT in the medical field during the Obama administration. The U.S. is a vast country, and the need for telemedicine is high. Subsequently, hospitals, which serve as hubs for digital health, have become smarter and the momentum is accelerating under the influence of COVID-19. China, Australia, and the United Kingdom are also making progress in digital health, with progress in these countries also led by government. Compared to these countries, Japan's digital healthcare is considered to be more than a decade behind. In a report issued in 2015, the OECD (Organization for Economic Cooperation and Development) ranked Japan last among OECD member countries in health information governance and utilization. The reasons for this are 1. ambiguity as to which government department is responsible for the digitization of healthcare; 2. the Galapagos syndrome affecting medical data handling and EHRs, which are also legacy systems; and 3. the impact of the "2000 problem" related to personal information protection legislation on personal health records (PHRs) [37], which prevents PHR standardization. Until a few years ago, Japan had sufficient technology for digitization, but the legal environment was not conducive to its use. However, the government has begun pursuing initiatives to overcome this impasse: In 2019, the Ministry of Health, Labor, and Welfare established the "Consortium for Accelerating AI Development in the Health and Medical Sector" and began government-led activities. In 2020, the Cabinet Office launched a project for social implementation of "Advanced Diagnosis and Treatment System by AI Hospitals," and as part of the project, the AI Hospital Promotion Center was established within the Japan Medical Association in 2021. And with the establishment of the Digital Agency and revision of the Personal Information Protection Law in 2022, standardization of PHRs and their widespread use are expected. The groundwork is now being laid for

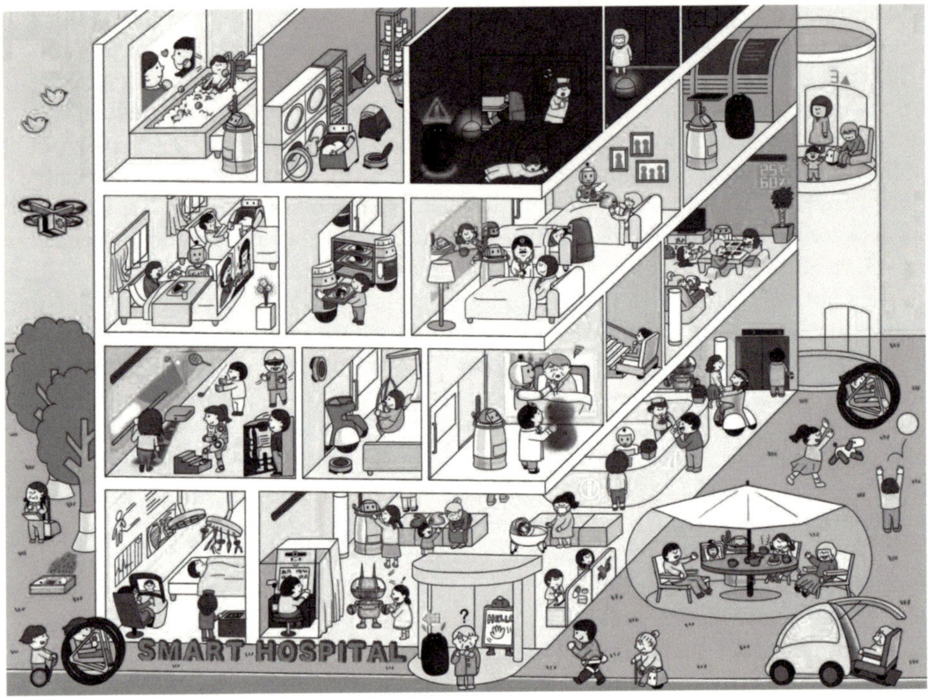

Figure 12.
The Smart Hospital Conceptual Diagram imagined by K. Terashima and M.Takahashi. A hospital of the future where people, IoT, AI, and robots can coexist in optimal symbiosis and perform efficient medical operations that are friendly to patients and healthcare professionals.

the standardization of digital health and smart hospitalization. **Figure 12** is the image of smart hospital concept by Kazuhiko Terashima, Toyohashi University of Technology, and Masami Takahashi, Masami Design Co., Ltd.

5. Future scope

An indispensable element to promote digital health and smart hospitalization is healthcare-industry collaboration. Until a few years ago, the healthcare sector was treated as off-limits, an inviolable sanctuary. However, we believe that the time has come to recognize that this attitude has hindered development of the healthcare sector and that change is overdue. Mutual understanding is essential for smooth healthcare-industry collaboration, and this requires the recruitment and training of human resources capable of working across the boundaries of medicine and engineering. Currently, we consider it necessary to have personnel who can translate needs in the medical sector, including those at the bedside, into the language of engineering and communicate them to those involved in development work and at the benchside [38–40]. In the future, we would like to see healthcare professionals who can do their own programming and engineering professionals who can serve as leaders of medical engineering in the medical field, and we would like to see the emergence of proactive personnel with skills and perspectives who are not content with the status quo. The Toyohashi University of Technology and the Smart Hospital Joint Research Chair were established, and research has begun on: 1. the application of speech recognition and natural language processing to electronic medical records; 2. image diagnosis support using AI; and 3. the application of behavior recognition technology in the medical field. We would like to provide useful digital health services to patients and the entire community by combining the data we have accumulated in our clinic with the university's advanced technology. Through our activities, we hope to develop a cadre of consummate professionals and invigorate the digital healthcare industry.

6. Conclusion

This paper began by introducing the research on next-generation nursing care systems and stations that have been developed by the authors' group, which are closely related to smart hospitals. Pursuing human-robot symbiosis, the authors' group has developed an assistive robot, which is not fully automated but follows human intention by distinguishing those tasks that are better performed by a human and has also developed a robot that is easy to operate, safe and secure, and comfortable to use. In addition to introducing those robots, the paper introduced care scheduling as part of a total system and referred to a showcase for care stations. Next, the paper presented an overview of smart hospitals and points to keep in mind as the project progresses. We believe it is important to deepen healthcare-industry collaboration and to train people capable of fulfilling important roles in the digital healthcare industry. We hope our work will stimulate future research and development of smart hospitals.

Acknowledgements

For valuable comment and support, we would like to thank Mr. Takahiko Suzuki, President, and Director of Medical Corporation Seishinkai, Toyohashi Heart Center; Mrs. Tatsuko Yamamoto, President of Social Welfare Corporation, Tenryu Kosei-kai; Mr. Makoto Fukihara, Senior Advisor of Sintokogio, Ltd.; and Dr. Norihide Kitaoka, Professor of Computer Science and Engineering, Toyohashi University of Technology, who heads Collaborative Smart Hospital Research by Toyohashi University of Technology and Toyohashi Heart Center.

Author details

Kazuhiko Terashima[1*], Kazuhiro Funato[2] and Takuyuki Komoda[3]

1 Toyohashi University of Technology, Toyohashi, Aichi Pref, Japan

2 KER Co. Ltd., Toyokawa, Aichi Pref, Japan

3 Toyohashi Heart Center, Medical Corporation Choushin-Kai, Toyohashi, Aichi Pref, Japan

*Address all correspondence to: terasima@tut.ac.jp

References

[1] Tsuruho Y. Cyber-physical system research for realization of super-smart society (in Japanese). Research, Technology and Plan. 2017;**32**(3): 218-219. DOI: 10.20801/jsrpim.32.3_218

[2] Gershenfeld N, Krikorian R, Cohen D. The internet of things. Scientific American. 2004;**291**(4):76-81. DOI: 10.1038/scientificamerican1004-76

[3] Xiaolong J, Benjamin W, Xueqi C, Yuanzhuo W. Significance and challenges of big data research. Big Data Research. 2015;**2**(2):59-64. DOI: 10.1016/j.bdr.2015.01.006

[4] Yamada N, Takashima Y, Kimura Y. Toward the realization of a super-smart society (Society 5.0): CPS/IoT and beyond (in Japanese). Information Management. 2017;**60**(5):325-334. DOI: 10.1241/johokanri.60.325

[5] Dimitrios S. The cyber-physical systems revolution. Computer. 2018; **51**(3):70-73. DOI: 10.1109/ MC.2018.1731058

[6] Hashimoto S. Smart house: A new living environment created by information and mechatronics technologies (in Japanese). Transactions of the Institute of Systems, Control and Information Engineers. 2003;**47**(3):108-112. DOI: 10.11509/ isciesci.47.3_108

[7] Ochahouse(in Japanese), Available from: http://is.ocha.ac.jp/siio/index.php?OchaHouse

[8] Nishida Y, Hori T, Suehiro T, Hirai S. Sensorized environment for self-communication based on observation of daily human behavior, Proceedings. In: IEEE/RSJ International Conference on Intelligent Robots and Systems (IROS 2000). Vol. 2. 2000. pp. 1364-1372. DOI: 10.1109/IROS.2000.893211

[9] Hirai S, Ueda H. Xi Home (Kusui-Home), An experimental house for living in Kyoto Sangyo University, IEICE Technical Report. MVE, Multimedia and Virtual Environment Basics, The Institute of Electronics, Information and Communication Engineers. 2010; **110**(35):43-50, Available from: https://cir.nii.ac.jp/crid/1571698601960198656

[10] Terashima K. Lifestyle support (medical, welfare, nursing care, rehabilitation) robot technology required by the super-aging society of the future (in Japanese). Joho Kiko. 2011:622

[11] Tasaki R, Kitazaki M, Miura J, Fukushima T, Terashima K. Design and development of medical care supporting robot (in Japanese). Journal of the Robotics Society of Japan. 2017;**35**(3): 1-9. DOI: 10.7210/jrsj.35.249

[12] Terashima K, Takenoshita S, Miura J, Tasaki R, Kitazaki M, Saegusa R, et al. Medical round robot – Terapio. Journal of Robotics and Mechatronics. 2014; **26**(1):112-114. DOI: 10.20965/jrm.2014. p0112

[13] Funato K, Sakurai H, Tasaki R, Terashima K. Development and experimental verification of a person tracking system of mobile robots using sensor fusion of inertial measurement unit and laser range finder for occlusion avoidance. Journal of Robotics and Mechatronics. 2021;**33**(1):33-43. DOI: 10.20965/jrm.2021.p0033

[14] Yamashita J, Asama H, Arai T, Ohta J, Kaneko T. A survey on trends of mobile robot mechanisms. Journal of the Robotics Society of Japan. 2003;**21**(3): 282-292. DOI: 10.7210/jrsj.21.282

[15] Ueno Y, Kitagawa H, Kakihara K, Terashima K. Development of the differential drive steering system using spur gear for omni-directional mobile robot (in Japanese). Proceedings of the Japan Society of Mechanical Engineers (Ed. C). 2012;**78**(789):1872-1885. DOI: 10.1299/kikaic.78.1872

[16] Ueno Y, Ohno T, Terashima K, Kitagawa H. The development of driving system with differential drive steering system for omni-directional mobile robot. International Conference on Mechatronics and Automation. 2009; **2009**:1089-1094. DOI: 10.1109/ICMA.2009.5246142

[17] Funato K, Tasaki R, Miyoshi T, Kakihara K, Terashima K. Design and analysis of novel nursing transformative assistive robot comprised of transfer and omnidirectional carrying (in Japanese). Journal of the Robotics Society of Japan. 2019;**37**(1):81-91. DOI: 10.7210/jrsj.37.81

[18] Funato K, Kenmotsu Y, Tasaki R, Sakakibara T, Kakihara K, Terashima K. Experimental analysis and anti-sway control of Jigiri behavior in a nursing lift. International Journal of Automation Technology. 2020;**14**(4):615-624. DOI: 10.20965/ijat.2020.p0615

[19] Matsuo K, Ochi Y, Tasaki R, Terashima K, Yamamoto T, Sakakibara T. Development of omni-directional body weight supported mobile walker which promotes active gait training(in Japanese). The Japan Society of Mechanical Engineers. 2012:172-175. DOI: 10.1299/jsmeshd.2012.172

[20] Mizushiri Y, Suzuki Y, Tasaki R, Miyoshi T, Kitagawa H, Terashima K. A novel mobile bodyweight-supported gait-training platform with systems for omnidirectional tracking and floor-reaction-force control. Journal Nursing Science and Engineering. 2017;**4**(2):121-132. DOI: 10.24462/jnse.4.2_121

[21] Tani Y, Tasaki R, Terashima K. Walking tracking system based on estimation of human posture in omni–directional mobile walker(in Japanese). Journal of the Robotics Society of Japan. 2019;**37**(2):161-167. DOI: 10.7210/jrsj.37.161

[22] Terashima K, Saegusa R. Robot-assisted gait training for older adults, NILTWAMOR and Lucia. Ageing and Digital Technology. Springer; 2019: 267-283. DOI: 10.1007/978-981-13-3693-5_16

[23] Saegusa R, Ito H, Duc MD. Human-care rounds robot with contactless breathing measurement. International Conference on Robotics and Automation (ICRA). 2019:6172-6177. DOI: 10.1109/ICRA.2019.8794037

[24] Terashima K, Watanabe K, Ueno Y, Masui Y. Auto-tuning control of power assist system based on the estimation of operator's skill level for forward and backward driving of omni-directional wheelchair. IEEE/RSJ International Conference on Intelligent Robots and Systems. 2010:6046-6051. DOI: 10.1109/IROS.2010.5652360

[25] Ueno Y, Kitagawa H, Kakihara K, Terashima K. Design and control for collision avoidance of power-assisted omni-directional mobile wheelchair system. IEEE/SICE International Symposium on System Integration (SII). 2011;**2011**:902-907. DOI: 10.1109/SII.2011.6147569

[26] Ueno Y, Kitagawa H, Kakihara K, Sakakibara T, Terashima K. Development of an innovative power-assist omni-directional mobile bed considering operator's characteristics. International Journal of Automation

Technology. 2014;**8**(3):490-499. DOI: 10.20965/ijat.2014.p0490

[27] Sata K, Yamagiwa K, Okada M. Social displaying with weakness for human-dependent sociable trash box(in Japanese). The Transactions of Human Interface Society. 2016;**18**(3):219-228. DOI: 10.11184/his.18.3_219

[28] Nishiwaki Y, Itashiki S, Okada M. Cooperative interactions generated by incompleteness in robot's utterance. The Transactions of Human Interface Society. 2019;**21**(1):1-12. DOI: 10.11184/his.21.1_1

[29] Okada M. Weak robots that make others their allies(in japanese). Transactions of the Institute of Systems, Control and Information Engineers. 2019;**63**(6):229-234. DOI: 10.11509/isciesci.63.6_229

[30] Sakaguchi T, Terashima K, Fukihara S, Scheduling device, scheduling method, and scheduling program, 2020, Patent publication: 2020-160785, Filing Date: 2019.3.26

[31] Sakaguchi T, Ishii R, Shirasuna M, Uchiyama N. Environment-adaptive genetic algorithm-based nesting scheduling for sheet-metal processing. Transactions of the Institute of Systems, Control and Information Engineers. 2020;**33**(2):39-48. DOI: 10.5687/iscie.33.39

[32] Terashima K. Toyohashi university of technology: The direction the center for human-robot symbiosis research should take and its achievements to date. Journal of Robotics and Mechatronics. 2021;**33**(1):6-10. DOI: 10.20965/jrm.2021.p0006

[33] Celia NA, Max P, Pascal G, Marco E, Andreas H. A survey of robotic systems for nursing care. Frontiers in Robotics and AI. 2022. DOI: 10.3389/frobt.2022.832248

[34] Lei Y, Yang L, XioaJuan Z. Smart hospital based on internet of things. Journal of Networks. 2012;7(10): 1654-1661. DOI: 10.4304/jnw.7.10.1654-1661

[35] Ko S, Jinzaki M, Kitagawa Y. Cabinet office AI hospital project at Keio university Hospital: Aspects of medical science(in Japanese). Ishiyaku Publishers. 2020;**274**(9):865-870

[36] Roberto M, Donato M. Healthcare digitalization and pay-for-performance incentives in smart hospital project financing. International Journal of Environmental Research and Public Health. 2020;**17**(7):2318. DOI: 10.3390/ijerph17072318

[37] Sugiyama H, Ikeda S, Muto M. A systematic review on definitions of Personal Health Record (PHR) in Japan (in japanese). Journal of the International University of Health and Welfare. 2012;**17**(22):20-31 Available from: https://cir.nii.ac.jp/crid/1050282677779345152

[38] Dairazalia S, Monica T. Jesús, activity recognition for the smart hospital. IEEE Intelligent Systems. 2008; **23**(2):50-57. DOI: 10.1109/MIS.2008.18

[39] Kashiwa K. Role and attitude of healthcare professionals in medical-industrial collaboration(in Japanese). Therapeutics & Engineering. 2021;**33**(1): 3-7

[40] Banu ÇU, Ertuğ O, Erkan D. Analysis of factors affecting IoT-based smart hospital design. Journal of Cloud Computing. 2020;**9**:67. DOI: 10.1186/s13677-020-00215-5

Chapter 9

Perspective Chapter: Digital Inclusion of the Farming Sector Using Drone Technology

Suman Dutta, Ajit Kumar Singh, Bhabani Prasad Mondal, Debashis Paul and Kiranmoy Patra

Abstract

Agriculture continues to be the primary source of income for most rural people in the developing economy. The world's economy is also strongly reliant on agricultural products, which accounts for a large number of its exports. Despite its growing importance, agriculture is still lagging behind to meet the demands due to crop failure caused by bad weather conditions and unmanaged insect problems. As a result, the quality and quantity of agricultural products are occasionally affected to reduce the farm income. Crop failure could be predicted ahead of time and preventative measures could be taken through a combination of conventional farming practices with contemporary technologies such as agri-drones to address the difficulties plaguing the agricultural sectors. Drones are actually unmanned aerial vehicles that are used for imaging, soil and crop surveillance, and a variety of other purposes in agricultural sectors. Drone technology is now becoming an emerging technology for large-scale applications in agriculture. Although the technology is still in its infancy in developing nations, numerous research and businesses are working to make it easily accessible to the farming community to boost the agricultural productivity.

Keywords: agriculture, crop productivity, drone technology, unmanned aerial vehicles, farm income

1. Introduction

Intensive agriculture has a number of detrimental environmental consequences. It contributes significant amounts of nitrogen and phosphorus to terrestrial ecosystems [1]. Agricultural chemicals pouring into adjacent water are also contributing to increased contamination of water bodies, as well as damage to water-related ecosystems [2]. Excessive fertilizer application can pollute the environment, but insufficient fertilizer application to replenish nitrogen and phosphorus lost via intense cropping can cause deterioration of soil fertility [3]. Furthermore, substantial soil deterioration is posing the ecosystem to reduce the productivity of many soils [4]. In addition to the environmental effect, the health risks associated with chemical usage in agriculture

Figure 1.
Interaction between drones and farmers to double farming income.

must be taken into account. Agricultural workers, their families, the residents of the areas surrounding crops, and farming sites may be at risk from toxins [5]. Furthermore, pesticides are absorbed by crops and natural resources posing a growing risk to animals and people, as well as having significant negative public health consequences [6]. These fatal consequences can be reduced by autonomous precision agriculture where chemicals like fertilizers and insecticides are only applied in the places of need rather than blanket application. Water deficiency, nutritional stress, and disease may be localized to assess, and a choice can be taken to solve the problem. In this regard, drones have been used in agriculture for large-area inspection, smart targeted irrigation, and fertilization [7, 8]. Some of the current uses of agriculture drones are monitoring of crop biomass, crop growth, food quality monitoring, and application in precision farming, such as determining the density of weed populations for site-specific herbicide treatments and logistic optimization of resources [9]. A few technical instances of drone application in precision farming are crop monitoring [10], pesticide application [11], soil and field analysis [12], and crop height estimations [13]. Special camera systems may gather excellent information from an unseen section of the electromagnetic spectrum termed Near-Infrared (NIR) to confirm the presence of algae in rivers or oil spills near coastlines [14]. All of these applications need the processing of pictures captured by a drone-mounted camera. The ability to detect regions where heavy irrigation is required or where a foliar disease is growing using a drone-mounted infrared camera can assist agronomists to save time, water, and agrochemicals. Simultaneously, better agricultural practices may result in higher crop output and productivity to enhance farm income. Recent years have seen a surge in interest in Human-Robot Interaction (HRI) from the academic community, government laboratories, IT companies, and the media [13]. Due to this interest, it would be beneficial to provide a general overview of HRI to act as a primer for individuals outside the field and to promote discussion on a common understanding of HRI inside the field. The body of literature is rapidly expanding with a focus on the technical disciplines of mechanical and electrical engineering, computer and control science, and artificial intelligence. The goals of drone usage are to provide a cogent overview of HRI-related concerns for emphasizing key themes on difficult topics that will probably have an immediate influence on the farming profession. The emphasis is on putting in place a multifaceted strategy to boost farmers' income by increasing productivity through the creation of resources (**Figure 1**).

2. Unmanned aerial vehicle

An unmanned aerial vehicle (UAV) can fly without a human pilot and is controlled by radio [15]. UAVs fill the void left by human mistakes and inefficiency in traditional farming practices. The goal of implementing drone technology is to eliminate any uncertainty or guessing and instead focus on accurate and dependable data [9]. Weather, soil conditions, and temperature are all important aspects of agriculture. Agriculture drones enable farmers to adapt unique circumstances and make thoughtful decisions. Multi rotors are a type of UAV that may be further defined by the number of rotors in its platform [5]. In the previous two decades, several types of UAV models such as Quad copter [16–18], Hexa copter [12, 13, 19], Octo copter [10, 20, 21], Fixed Wing [22–24], and Single Rotor Helicopter [11, 25–32] have been deployed. Fixed wing UAVs have a completely different design from multirotor. A single-rotor helicopter features only one large rotor on top and one little rotor on the UAV's tail. Quad copters, Hexa copters, and Octo copters are multi-rotors with four, six, or eight rotors, respectively that lift and propel them. A quad copter is a type of UAV with four rotors. These rotors are responsible for the quad copter's lift. The two opposing rotors rotate in a clockwise direction, while the other two rotate in a counter-clockwise direction. Pitch (backward and forward), roll (left and right), and yaw (clockwise and counter-clockwise) are the three modes of quad copter movement around the axis [5].

3. The architecture of a drone for precision agriculture

Drones are semi-autonomous in precision agriculture, as well as in disaster assistance, construction inspection, and traffic monitoring [9]. In that instance, the drone must follow the definition of a straight path in terms of waypoints and height. As a result, the drone must have a positioning measurement system such as the Global Navigation Satellite System (GNSS) on the board in order to determine its location in relation to the route points. It also has an altimeter such as barometer, laser altimeter, or ultrasonic sensor for flighting at consistent heights. The APM Planner is an example of software for determining the mission trajectory. According to Brzozowski et al. [33], the fundamental design of a drone comprises the following components: (a) frame, (b) brushless motors, (c) Electronic Speed Control (ESC) modules, (d) control board, (e) Inertial Navigation System (INS), and (f) transmitter and receiver modules. Multispectral, infrared, RGB (Red-Green-Blue), and Light Detection and Ranging (LiDAR) systems are among the sensors integrated into drones in precision agriculture [9]. The status of the observed vegetation is quantified using multispectral cameras in terms of chlorophyll content, leaf water content, Leaf Area Index (LAI), ground cover, and the Normalized Difference Vegetation Index (NDVI) [2]. Due to the elevated temperature of stressed plants, thermal cameras have shown a significant potential for detecting water stress in crops. A drone that uses thermal and multispectral sensors has the potential to monitor vegetation in a very less period of time [34]. To digitize the terrain surface and produce the Digital Terrain Model (DTM) or Digital Surface Model (DSM) of the monitored region, RGB cameras, and LiDAR systems are commonly used [35]. The DTM depicts the soil's ground level without taking into account plant height. The DSM, on the other hand, depicts the earth's surface and all of its inhabitants. The DSM and DTM are extrapolated using the commercial program Pix4Dmapper [35]. By subtracting the DTM from the DSM, the differential model of the vine rows is produced. The UAV is equipped with a number of components that allow it to regulate its mobility in response to the perceived surroundings (**Table 1**). A drone that may be used for precision agriculture must meet

Component	Function
Accelerometer	For quantity the acceleration the UAV
Air Pressure Sensor	Gases or liquids measurement
Altimeter	Elevation measurement
Anemometer	Wind Speed measurement
Barometer	To measure the atmospheric pressures
BLDC	To motion control
Camera (RGB)	To capture visual images
Digital Temperature	Temperature measurement
ESC	Regulation of the speed of BLDC
Filter papers	Fine substances separation
GPS	Offers geo location of an entity
Gyro	For rotational motion
Humidity indicator	To quantify the moisture in air
Hyper spectral camera	Images at narrow spectral bands
IMU	Angular rate and forces measurement
Laser scanner 2D	To captures shape of the entity
Magnetometer	Magnetic field measurement
Microsoft Kinect	For motion sensing
Multispectral Camera	Images at specific frequencies
PWM controller	For pulsing signal
Telemetry	To obtain live data from UAV
Thermal Camera	For recording low light imaginary
Video Camera	To capture electronic motion of the objects
Water sensitive paper	For spray coverage assessment
WSN	Sensing the environmental conditions

Table 1.
Hardware components of the drone.

the required capabilities namely (a) the drone must fly according to the waypoints, (b) the drone must control its height of flying, (c) avoidance of obstacles during the flight, (d) the drone must automatically land based on the state of the battery, and (e) the acquired images must be stabilized [9]. Their hardware implementations are entirely dependent on important factors such as weight, range of flight, payload, configuration, and prices [5]. Hardware component assembly, software system integration, aerodynamic modeling, autonomous flight control, and implementations in the farming sectors are among the important ingredients required in developing a miniature autonomous unmanned rotorcraft aircraft [28].

4. Precision agriculture with the use of drones

Drones are UAVs that are utilized in a variety of businesses for monitoring [36]. Until now, they were largely employed by enterprises in the mining and construction industries, the army, and hobbyists. However, drone technology is now becoming more widely available for application in agriculture (**Figure 2**) [37]. For instance, Yamaha RMAX is a petrol-powered unmanned aerial vehicle created for pesticide spraying in Asian rice fields [26]. Despite the fact that the technology is still in its infancy in agriculture, numerous businesses are working to make it readily available to farmers to boost agricultural productivity. Agricultural drones can be used for soil and field studies to help in field planning. They can be used to mount sensors that measure soil moisture content, topographical conditions, soil conditions, soil erosion, soil nutrients, and soil fertility. For drones to successfully integrate into human environments, they must be dependable and secure. When utilized to aid humans physically, drones should reduce human stress and tiredness, boost human force, speed, and precision, and overall enhance the quality of farming life. Humans, on the other hand, may provide expertise and comprehension to make sure that tasks are completed appropriately.

4.1 Application of drones using multispectral and thermal cameras

It is possible to estimate chlorophyll absorption, water deficiency, pesticide absorption, diseases, and nutritional stress in precision agriculture using reflectance measurements [9]. Each pixel in the generated image has a sampled spectral measurement of the reflectance, which may be read to identify the substance present in the scene after post-processing [37]. A crop monitoring system for pesticide spraying with UAV comprises of an automated drone and a sprinkling system with a multispectral camera [5]. The sprinkling system is linked to the UAV's lower section, with a nozzle beneath the pesticide tank that sprays the pesticide downstream. The multi spectral camera scans the whole crop field and provides a spatial map as the first method of monitoring. This map shows the status of the crop using NDVI, and the farmer then decides which herbicides and fertilizers are required to use on the crop. The acquisition of spectral image data involves four resampling operations or calibrations namely spatial, spectral, radiometric, and temporal [9]. The Ground Sample Distance (GSD) refers to the spatial sampling in which the distance in meters is measured on the ground between two successive pixel centers. It is determined by the sensor aperture and the height of the flight. In contrast, in spectral sampling, decomposing the radiance obtained in each spatial pixel into a finite number of wavebands is carried out. In radiometric sampling, the resolution of the Analog to Digital Converter (ADC) used to sample the radiance obtained in each spectral channel

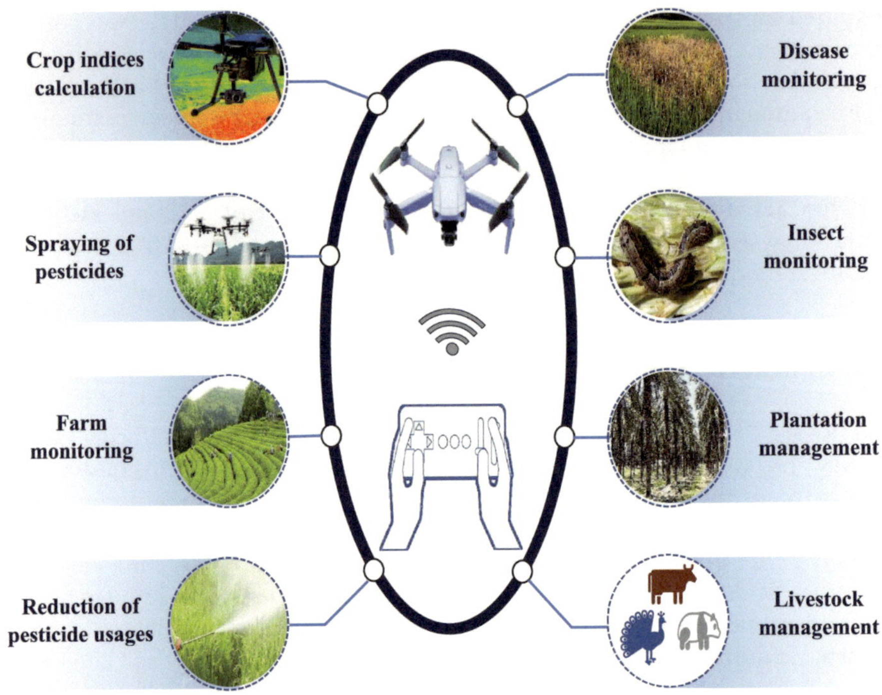

Figure 2.
Applications of the drone technology in agriculture.

correlates to the radiometric resolution. Temporal sampling, on the other hand, refers to the collection of several spectral photographs of the same scene at various times. These four sample processes must be considered while designing a flight operation using a multispectral camera and drone platform. The photos were taken by the drone offer data related to the radiance in each pixel. Image processing algorithms are used while measuring reflectance to adjust for effects owing to atmospheric absorption and the spectrum of solar light. Drones are a better platform for precision agricultural scanning than satellites because of the high spatial resolution [37]. Satellites equipped with multispectral and thermal cameras survey the large area for agriculture. Drones provide far more flexibility than satellites. In a ground-based setting, the drone multispectral and thermal sensors sample spectral wavebands concurrently across a vast region [36].

4.2 Application of drones using RGB cameras

The pictures captured by drones equipped with RGB cameras are utilized in precision agriculture to extrapolate DTM and DSM for the examined region. The right mission parameters must be determined to achieve the goal based on the spatial resolution and measurement accuracy of the rebuilt DTM and DSM [9]. The spatial resolution of multispectral and thermal cameras is defined in terms of GSD. The camera resolution and the flight height are determined based on the GSD. By capturing two successive photographs from the camera at two different waypoints, the height measurements of the landscape and the objects in the scene could be determined [38]. In most cases, a 70 percent overlapping factor between the two photos is used. The precision of a 3D reconstruction is mostly determined by the quality of the drone position measuring system. Several techniques for localizing the drone during the flight can be used: (a) differential GPS systems with a position accuracy in the order of 1 m, (b) real-time kinematic (RTK) GPS with an accuracy in the order of 2 cm, and (c) simultaneous localization and mapping (SLAM) based techniques with a position accuracy in the order of 10 cm [9].

4.3 Monitoring of the crop plants

Farmers deal with a variety of issues, such as the high cost of labor, health issues caused by contact with chemicals (fertilizers and insecticides) when using them in the field, bug or animal bites, etc. In this situation, drones can aid farmers by assisting them in avoiding these issues in addition to the advantages of being a green technology. In a single trip, the UAVs may cover hundreds of hectares of land to monitor the crop status using a variety of sensors [39]. Thermal and multi-spectral cameras were used to capture the reflectance of the plant canopy, which is placed on the quad copter's backside [10, 37]. The camera takes one image per second, saves it in memory, and transmits it to the ground station via telemetry using wireless Mavlink protocol. Blue (440–510 nm), green (520–590 nm), red (630–685 nm), red-edge (690–730 nm), and near-infrared (760–850 nm) wavelength were used to take the photos [5]. The Geographic indicator NDVI was used to examine data from the multispectral camera through telemetry [40, 41]. The embedded GPS module saves the GPS coordinates of every captured photo. The GPS coordinates are then kept in the UAV, allowing pesticides to be sprayed automatically without human intervention.

4.4 Drone to provide sprinkling system

The sprinkling system is usually linked to the bottom portion of the UAV, with a nozzle beneath the pesticide tank to spray the pesticide downstream [5]. The sprinkler system is made up of two modules namely the sprinkler system itself and the controller. Spraying substances such as pesticides or fertilizers and a spraying nozzle are included in the sprinkling system, whereas, a controller is used to turn on the nozzle for spraying. Different spraying rates and nozzles used in UAVs for spraying have been analyzed. Different Nozzle Type includes Flat fan [19], Micron air-A+ [29], Micron air ULV-A+ (ultra-low volume) [11], Universal [42], Centrifugal [27], Fan-shaped (electrostatic) [43], Electric centrifugal [31], Rotary atomizer [32] and conical [21]. A pressure pump in the sprinkler system exerts pressure to flow out the insecticide through the nozzle. It is necessary to have a motor driver integrated circuit to adjust the pressure of the pump as per requirement.

4.5 Miscellaneous application of the drone

Pollination has also been accomplished utilizing wind energy provided by UAVs. Jiyu et al. [44] investigated how the dispersal of rice pollen was affected by helicopter-type UAV wind power. They discovered that the UAV-created wind field had an uneven effect on pollen dispersal. As the insect infestation spreads fast, early detection is critical to avoid huge economic losses. A combination of high-resolution RGB cameras and multi-spectrum sensors can be placed on UAVs to evaluate the infestation intensity in potato fields [45]. Using high-quality spectral data, they demonstrated accurate and rapid pathogen detection. UAVs outfitted with multispectral cameras and thermal sensors may also identify locations where water is scarce [46]. Baluja et al. [47] conducted research to collect data for water management and irrigation control utilizing thermal and multispectral imagery. To optimize irrigation benefits, the scientists experimented with numerous UAVs. Irrigation automation will be deployed efficiently via a collaborative system combining UAVs in future smart farming. It is not a matter of surprise that UAVs can make planting more efficient [48]. When planting seeds and nutrients, a method must employ them to disperse evenly to offer ideal circumstances for plant growth and development. Although the use of UAVs for planting is still in its early stages, it is believed that this approach would yield efficient results if the UAV is equipped with image recognition technology and planting tasks [2]. UAVs are typically outfitted with cameras and sensors for crop monitoring for spraying pesticides. Yamaha RMAX, an unmanned helicopter, was launched for pest control and crop monitoring applications in agriculture [26]. Various UAV types have already been used for military and civilian purposes [5]. Using laser power beaming technology, UAVs can be used to extend the flying duration [49]. The proportional integral derivative (PID) controlling method is used to regulate the aerodynamic domain, tuning, and trimming phases of the UAV [50–52]. The images were then processed and analyzed using the NDVI method. Sensors and vision systems can also help UAVs to reach their full potential [53]. Another technique was implemented on the ground, in which a sprayer system was put on a UAV to spray pesticides [36]. The combination of UAVs with a sprayer system has the potential to provide a platform for pest management and vector control. Heavy lift UAVs are necessary for large-area spraying for this purpose [20]. In pesticide applications, the PWM (pulse width modulation) controller improves the effectiveness of the spraying system placed on the UAV [11, 29]. Pesticide deposition from the developed UAV is virtually identical to that of ground-based sprayers. The RMAX is a crop sprayer designed for high-value crops [5]. Because of their speed and precision, UAVs are increasingly being used in spraying operations. However, several variables affect the spraying operation in the field such as some sections in the crop field not being fully covered during spraying, overlapping spraying in the crop areas, and the spraying process extending beyond the outer margins of the crop field [5]. To overcome these obstacles, a swarm of UAVs was utilized in a control loop of an algorithm for efficient spraying operations [54]. In a lower altitude context, a blimp integrated quad copter aerial automated pesticide sprayer (AAPS) was designed for pesticide spraying based on GPS coordinates [18]. A low-cost, user-friendly pesticide spraying drone operated by an Android app called 'Freyr' was developed [55]. Using a double pulsed laser, a particle image velocimetry approach was employed to quantify the downwash flow field droplet mobility and deposition over the crop at various spinning speeds of the rotors of an octa copter [21]. Furthermore, filter papers and water-sensitive papers are employed to investigate spraying deposition and droplet coverage over many spraying swaths [30, 56]. However, numerous issues need to be resolved in order to have dependable and safe physical human-drone contact. In the near future, it will be important to successfully integrate drones into everyday situations; as a result, dependability and safety standards must be established. Although there are undoubtedly additional cognitive challenges involved due to how humans perceive drones (and vice versa), as well as other objective measures linked to defect detection and isolation. Safety and reliability in particular serve as the primary evaluation criteria for the mechanical design of control systems. Focus is placed on dependability with special attention paid to sensors, control architectures, failure management, and fault tolerance.

5. Drones as vehicles to increase farmer income: a perspective from Indian initiatives

The Union Ministry of Agriculture and Farmers Welfare has published rules in the year 2021 to make drone technology accessible to stakeholders to provide a significant boost to the promotion of precision farming in India [57]. The "Sub-Mission on Agricultural Mechanization" (SMAM) guidelines have been updated, and now allow for grants of up to 100% of the cost of an agricultural drone in order to purchase drones for use in large-scale field tests. Farmers can make decisions to increase crop yield since they have access to timely information and advice through the internet and telecom mediums like the Kisan Call Center and Kisan Suvidha Application. Crop health is closely monitored in real-time using drones and artificial intelligence. The drone uses include mapping water spread areas, water sampling, mapping macrophyte infestation, and aquaculture management techniques [57]. Variable rate technology is also employed for pesticide and liquid fertilizer applications. Precision livestock farming also makes use of artificial intelligence and drone technologies, notably for monitoring the health of cattle. The focus is on implementing a multi-faceted approach to increase farmers' income by raising production through the development of resources for better irrigation, using nutrient inputs effectively, lowering post-harvest losses, adding value, reforming agriculture marketing, reducing risk, and offering security and assistance [57].

6. Limitation of drone technology

Agriculture with intelligent practices may be used anywhere around the globe. In developing and under-developing nations, smart agriculture using UAVs has begun to increase the farm output in agriculturally important crops. Furthermore, it allows farmers to produce more from their smaller holdings [58, 59]. However, there are considerable limitations and issues in their application at this early level of research and development. Self-life of battery and flight time restrictions are the big issues for the potential application of UAVs [2]. Research on increasing the power of drone batteries is still going on to find a good solution. Lithium-ion batteries are now in use as they provide a bigger capacity and longer flying time than traditional batteries. Despite the fact that battery management necessitates regular maintenance, the majority of UAV operators result in higher expenditures as the UAV necessitates a more frequent replacement of the batteries. As a result, researchers are creating optimal hybrid battery solutions [60, 61]. Swarm-control approaches, which employ numerous UAVs to efficiently accomplish a variety of tasks, are also being investigated by researchers [2]. Swarms provide realistic methods for reducing battery costs and operating more efficiently with shorter flight hours. Multimodal input, such as visual, tactile, and aural feedback, is also needed to improve the user interface [2]. UAVs are often difficult to operate by ordinary people in agriculture and therefore, are typically used by specialists to do agricultural operations. Improving the user interface can make it easier for people who are unfamiliar with drones to control them. Human-centered user interface and feedback are very effective when dealing with multi-UAV systems [62]. To use drone technology, farms need to get approval from the government authorities and eventually restrict its use. Even though agricultural UAVs have enormous promise in agriculture, at these early stages of study and development, there are significant restrictions and problems to use them. HRI with automation has received great attention from various communities to date, particularly in the farming sector. This is in line with the fact that HRI in aircraft piloting has long been a topic of active study in the field of human intelligence [63]. In any case, research into many facets of human interaction as well as participation in the creation of drones is very important. While those in the human factors field can undoubtedly benefit from a deeper comprehension of dynamic control of computer science guided by artificial intelligence [64]. Therefore, there should be a very close look for opportunities to collaborate on research, conceptual design, and evaluation with engineers in these fields.

7. Conclusion

Intensive agriculture has a lot of negative effects on the environment. It supplies large quantities of nitrogen and phosphorus to terrestrial ecosystems leading to growing pollution of water courses and bodies of water, as well as ecological harm. Furthermore, significant soil degradation is reducing the productivity of many soils, putting the ecosystem at risk. Aside from the environmental impact, the health hazards linked with agricultural chemical use must be considered. Pesticides are also absorbed by crops and natural resources, creating a rising threat to public health. Chemicals like fertilizer and pesticides are only sprayed where they are needed rather than being distributed across a large region. Therefore, the implications of autonomous precision farming using a drone minimizes such negative impacts. UAVs are already being used in precision agriculture and remote sensing in developed nations. It is extremely quick, and it has the potential to minimize a farmer's workload. However, precision agriculture with UAVs is still in its early stages, with room for advancement in both technology and agriculture applications. A large number of experimental investigations using UAV-based remote sensing for agricultural applications have been conducted. Drones have been employed in agriculture for large-area surveillance and smart targeted irrigation and fertilization. Using a drone with an infrared camera to detect areas where high irrigation is necessary or where a foliar disease is spreading can help agronomists save time, water, and usage of agrochemical inputs. Better agricultural operations using advanced technology like drones may result in improvement in crop yield and productivity.

Acknowledgements

The participant is also thankful to the Human Resource Development Group (HRDG) division of the Council of Scientific and Industrial Research (CSIR), New Delhi, India for the Junior Research Fellowship (File No.: 09/083(0383)/2019-EMR-I) to pursue his Ph.D. program. The participant also shows his sincere gratitude to Ajit Kumar Singh, Bhabani Prasad Mondal, Debashis Paul, and Kiranmoy Patra for reviewing and correcting the manuscript.

Conflict of interest

The authors declare no conflict of interest.

Author details

Suman Dutta[1*], Ajit Kumar Singh[2], Bhabani Prasad Mondal[3], Debashis Paul[4] and Kiranmoy Patra[5]

1 Division of Genetics, ICAR-Indian Agricultural Research Institute, India

2 Division of Food Science and Post-Harvest Technology, ICAR-Indian Agricultural Research Institute, India

3 Division of Agricultural Physics, ICAR-Indian Agricultural Research Institute, India

4 ICAR-Central Institute for Cotton Research, India

5 Division of Agronomy, ICAR-Indian Agricultural Research Institute, India

*Address all correspondence to: duttasumansiliguri@gmail.com

References

[1] Vitousek PM, Mooney HA, Lubchenco J, Melillo JM. Human domination of Earth's ecosystems. Science. 1997;**277**(5325):494-499

[2] Kim J, Kim S, Ju C, Son HI. Unmanned aerial vehicles in agriculture: A review of perspective of platform, control, and applications. IEEE Access. 2019;**7**:105100-105115

[3] Tilman D, Cassman KG, Matson PA, Naylor R, Polasky S. Agricultural sustainability and intensive production practices. Nature. 2002;**418**(6898):671-677

[4] Salmelin J, Polonen I, Puupponen HH, HamalainenH, KarjalainenAK, VaisanenA, et al. Hyperspectral imaging of macroinvertebrates—a pilot study for detecting metal contamination in aquatic ecosystems. Water, Air, and Soil Pollution. 2018;**229**(9):1-13

[5] Mogili UR, Deepak BBVL. Review on application of drone systems in precision agriculture. Procedia Computer Science. 2018;**133**:502-509

[6] Igbedioh SO. Effects of agricultural pesticides on humans, animals, and higher plants in developing countries. Archives of Environmental Health: An International Journal. 1991;**46**(4):218-224

[7] Bacco M, Ferro E, Gotta A. UAVs in WSNs for agricultural applications: An analysis of the two-ray radio propagation model. In: Sensors. New York, USA: IEEE; 2014. pp. 130-133

[8] Gago J, Douthe C, Coopman RE, Gallego PP, Ribas-Carbo M, Flexas J, et al. UAVs challenge to assess water stress for sustainable agriculture. Agricultural Water Management. 2015;**153**:9-19

[9] Daponte P, De Vito L, Glielmo L, Iannelli L, Liuzza D, Picariello F, et al. A review on the use of drones for precision agriculture. In: IOP Conference Series: Earth and Environmental Science. Vol. 275. 2019. p. 012022

[10] Bendig J, Bolten A, Bareth G. Introducing a low-cost mini-UAV for thermal-and multispectral-imaging. The International Archives of the Photogrammetry, Remote Sensing and Spatial Information Sciences. 2012;**39**:345-349

[11] Huang Y, Hoffmann WC, Lan Y, Wu W, Fritz BK. Development of a spray system for an unmanned aerial vehicle platform. Applied Engineering in Agriculture. 2009;**25**(6):803-809

[12] Primicerio J, Di Gennaro SF, Fiorillo E, Genesio L, Lugato E, Matese A, et al. A flexible unmanned aerial vehicle for precision agriculture. Precision Agriculture. 2012;**13**(4):517-523

[13] Anthony D, Elbaum S, Lorenz A, Detweiler C. On crop height estimation with UAVs. In: 2014 IEEE/RSJ International Conference on Intelligent Robots and Systems. Chicago, IL, USA: IEEE; 2014. pp. 4805-4812

[14] Manfredonia I, Stallo C, Ruggieri M, Massari G, Barbante S. An early-warning aerospace system for relevant water bodies monitoring. In: 2015 IEEE Metrology for Aerospace (MetroAeroSpace). 2015. pp. 536-540

[15] Everaerts J. The use of unmanned aerial vehicles (UAVs) for remote sensing and mapping. The International Archives of the Photogrammetry, Remote Sensing and Spatial Information Sciences. 2008;**37**(2008):1187-1192

[16] Kabra TS, Kardile AV, Deeksha MG, Mane DB, Bhosale PR, Belekar AM. Design, development and optimization of a quad-copter for agricultural applications. International Research Journal of Engineering and Technology (IRJET). 2017;**4**(7):1632-1636

[17] Patel PN, Patel MA, Faldu RM, Dave YR. Quadcopter for agricultural surveillance. Advance in Electronic and Electric Engineering. 2013;**3**(4):427-432

[18] Vardhan PH, Dheepak S, Aditya PT, Arul S. Development of automated aerial pesticide sprayer. International Journal of Engineering Science and Research Technology. 2014;**3**(4):458-462

[19] Yallappa D, Veerangouda M, Maski D, Palled V, Bheemanna M. Development and evaluation of drone mounted sprayer for pesticide applications to crops. In: 2017 IEEE Global Humanitarian Technology Conference (GHTC). San Jose, CA, USA: IEEE; 2017. pp. 1-7

[20] Sarghini F, De Vivo A. Analysis of preliminary design requirements of a heavy lift multirotor drone for agricultural use. Chemical Engineering Transactions. 2017;**58**:625-630

[21] Qing T, Ruirui Z, Liping C, Min X, Tongchuan Y, Bin Z. Droplets movement and deposition of an eight-rotor agricultural UAV in downwash flow field. International Journal of Agricultural and Biological Engineering. 2017;**10**(3):47-56

[22] Herwitz S, Johnson L, Arvesen J, Higgins R, Leung J, Dunagan S. Precision agriculture as a commercial application for solar-powered unmanned aerial vehicles. In: 1st UAV Conference. Portsmouth, Virginia: AIAA; 2002. p. 3404

[23] Herwitz SR, Johnson LF, Dunagan SE, Higgins RG, Sullivan DV, Zheng J, et al. Imaging from an unmanned aerial vehicle: Agricultural surveillance and decision support. Computers and Electronics in Agriculture. 2004;**44**(1):49-61

[24] Pederi YA, Cheporniuk HS. Unmanned aerial vehicles and new technological methods of monitoring and crop protection in precision agriculture. In: 2015 IEEE International Conference Actual Problems of Unmanned Aerial Vehicles Developments (APUAVD). Kyiv, Ukraine: IEEE; 2015. pp. 298-301

[25] Huang Y, Hoffman WC, Lan Y, Fritz BK, Thomson SJ. Development of a low-volume sprayer for an unmanned helicopter. Journal of Agricultural Science. 2015;**7**(1):148

[26] Giles D, Billing R. Deployment and performance of a UAV for crop spraying. Chemical Engineering Transactions. 2015;**44**:307-312

[27] Xue X, Lan Y, Sun Z, Chang C, Hoffmann WC. Develop an unmanned aerial vehicle based automatic aerial spraying system. Computers and Electronics in Agriculture. 2016;**128**:58-66

[28] Cai G, Chen BM, Lee TH. An overview on development of miniature unmanned rotorcraft systems. Frontiers of Electrical and Electronic Engineering in China. 2010;**5**(1):1-14

[29] Zhu H, Lan Y, Wu W, Hoffmann WC, Huang Y, Xue X, et al. Development of a PWM precision spraying controller for unmanned aerial vehicles. Journal of Bionic Engineering. 2010;**7**(3):276-283

[30] Shilin W, Jianli S, Xiongkui H, Le S, Xiaonan W, Changling W, et al. Performances evaluation of four typical unmanned aerial vehicles used

for pesticide application in China. International Journal of Agricultural and Biological Engineering. 2017;**10**(4):22-31

[31] Xinyu X, Kang T, Weicai Q, Yubin L, Huihui Z. Drift and deposition of ultra-low altitude and low volume application in paddy field. International Journal of Agricultural and Biological Engineering. 2014;7(4):23-28

[32] Qin W, Xue X, Zhang S, Gu W, Wang B. Droplet deposition and efficiency of fungicides sprayed with small UAV against wheat powdery mildew. International Journal of Agricultural and Biological Engineering. 2018;**11**(2):27-32

[33] Brzozowski B, Daponte P, De Vito L, Lamonaca F, Picariello F, Pompetti M, et al. A remote-controlled platform for UAS testing. IEEE Aerospace and Electronic Systems Magazine. 2018;**33**(8):48-56

[34] Berni JA, Zarco-Tejada PJ, Suarez L, Fereres E. Thermal and narrowband multispectral remote sensing for vegetation monitoring from an unmanned aerial vehicle. IEEE Transactions on Geoscience and Remote Sensing. 2009;**47**(3):722-738

[35] Burgos S, Mota M, Noll D, Cannelle B. Use of very high-resolution airborne images to analyse 3D canopy architecture of a vineyard. The International Archives of Photogrammetry, Remote Sensing and Spatial Information Sciences. 2015;**40**(3):399

[36] Zhang C, Kovacs JM. The application of small unmanned aerial systems for precision agriculture: A review. Precision Agriculture. 2012;**13**(6):693-712

[37] Colomina I, Molina P. Unmanned aerial systems for photogrammetry and remote sensing: A review. ISPRS Journal of Photogrammetry and Remote Sensing. 2014;**92**:79-97

[38] Daponte P, De Vito L, Mazzilli G, Picariello F, Rapuano S. A height measurement uncertainty model for archaeological surveys by aerial photogrammetry. Measurement. 2017;**98**:192-198

[39] Simelli I, Tsagaris A. The use of unmanned aerial systems (UAS) in agriculture. Kavala, Greece: HAICTA; 2015. pp. 730-736

[40] Reinecke M, Prinsloo T. The influence of drone monitoring on crop health and harvest size. In: 2017 1st International Conference on Next Generation Computing Applications (NextComp). Mauritius: IEEE; 2017. pp. 5-10

[41] Bhandari AK, Kumar A, Singh GK. Feature extraction using normalized difference vegetation index (NDVI): A case study of Jabalpur city. Procedia Technology. 2012;**6**:612-621

[42] Meivel SM, Maguteeswaran R, Be NG, Srinivasan G. Quadcopter UAV based fertilizer and pesticide spraying system. Int. Acad. Res. J. Eng. Sci. 2016;**1**:8-12

[43] Yanliang Z, Qi L, Wei Z. Design and test of a six-rotor unmanned aerial vehicle (UAV) electrostatic spraying system for crop protection. International Journal of Agricultural and Biological Engineering. 2017;**10**(6):68-76

[44] Jiyu L, Lan Y, Jianwei W, Shengde C, Cong H, Qi L, et al. Distribution law of rice pollen in the wind field of small UAV. International Journal of Agricultural and Biological Engineering. 2017;**10**(4):32-40

[45] Nebiker S, Lack N, Abacherli M, Laderach S. Light-weight multispectral

UAV sensors and their capabilities for predicting grain yield and detecting plant diseases. In: International Archives of the Photogrammetry, Remote Sensing and Spatial Information Sciences. 2016;**XLI-B1**: 963-970

[46] Romero-Trigueros C, Nortes PA, Alarcon JJ, Hunink JE, Parra M, Contreras S, et al. Effects of saline reclaimed waters and deficit irrigation on citrus physiology assessed by UAV remote sensing. Agricultural Water Management. 2017;**183**:60-69

[47] Baluja J, Diago MP, Balda P, Zorer R, Meggio F, Morales F, et al. Assessment of vineyard water status variability by thermal and multispectral imagery using an unmanned aerial vehicle (UAV). Irrigation Science. 2012;**30**(6):511-522

[48] Salaan CJ, Tadakuma K, Okada Y, Sakai Y, Ohno K, Tadokoro S. Developmentandexperimentalvalidation of aerial vehicle with passive rotating shell on each rotor. IEEE Robotics and Automation Letters. 2019;**4**(3):2568-2575

[49] Achtelik MC, Stumpf J, Gurdan D, Doth KM. Design of a flexible high performance quadcopter platform breaking the MAV endurance record with laser power beaming. In: 2011 IEEE/RSJ International Conference on Intelligent Robots and Systems. San Francisco, CA, USA: IEEE; 2011. pp. 5166-5172

[50] Sohail S, Nasim S, Khan NH. Modeling, controlling and stability of UAV Quad Copter. In: 2017 International Conference on Innovations in Electrical Engineering and Computational Technologies (ICIEECT). Karachi, Pakistan: IEEE; 2017. pp. 1-8

[51] Qasim M, Susanto E, Wibowo AS. PID control for attitude stabilization of an unmanned aerial vehicle quad-copter. In: 2017 5th International Conference on Instrumentation, Control, and Automation (ICA). Yogyakarta, Indonesia: IEEE; 2017. pp. 109-114

[52] Mallick TC, Bhuyan MAI, Munna MS. Design and implementation of an UAV (Drone) with flight data record. In: 2016 International Conference on Innovations in Science, Engineering and Technology (ICISET). Dhaka, Bangladesh: IEEE; 2016. pp. 1-6

[53] Gupte S, Mohandas PIT, Conrad JM. A survey of quadrotor unmanned aerial vehicles. In: 2012 Proceedings of IEEE Southeastcon. Orlando, FL, USA: IEEE; 2012. pp. 1-6

[54] Yao L, Jiang Y, Zhiyao Z, Shuaishuai Y, Quan Q. A pesticide spraying mission assignment performed by multi-quadcopters and its simulation platform establishment. In 2016 IEEE Chinese Guidance, Navigation and Control Conference (CGNCC). Nanjing, China: IEEE; 2016. pp. 1980-1985

[55] Spoorthi S, Shadaksharappa B, Suraj S, Manasa VK. Freyr drone: Pesticide/fertilizers spraying drone-an agricultural approach. In: 2017 2nd International Conference on Computing and Communications Technologies (ICCCT). Chennai, India: IEEE; 2017. pp. 252-255

[56] Tang Y, Hou CJ, Luo SM, Lin JT, Yang Z, Huang WF. Effects of operation height and tree shape on droplet deposition in citrus trees using an unmanned aerial vehicle. Computers and Electronics in Agriculture. 2018;**148**:1-7

[57] https://agricoop.nic.in › files › SOP for Drone_1

[58] Zhong Y, Wang X, Xu Y, Wang S, Jia T, Hu X, et al. Mini-UAV-borne hyperspectral remote sensing:

From observation and processing to applications. IEEE Geoscience and Remote Sensing Magazine. 2018;**6**(4):46-62

[59] Wahab I, Hall O, Jirstrom M. Remote sensing of yields: Application of UAV imagery-derived ndvi for estimating maize vigor and yields in complex farming systems in sub-saharan africa. Drones. 2018;**2**(3):28

[60] Saha B, Koshimoto E, Quach CC, Hogge EF, Strom TH, Hill BL, et al. Battery health management system for electric UAVs. In: 2011 Aerospace Conference. Big Sky, MT, USA: IEEE; 2011. pp. 1-9

[61] Lee B, Park P, Kim C, Yang S, Ahn S. Power managements of a hybrid electric propulsion system for UAVs. Journal of Mechanical Science and Technology. 2012;**26**(8):2291-2299

[62] Hong A, Lee DG, Bulthoff HH, Son HI. Multimodal feedback for teleoperation of multiple mobile robots in an outdoor environment. Journal on Multimodal User Interfaces. 2017;**11**(1):67-80

[63] Sheridan TB. Human–robot interaction: Status and challenges. Human Factors. 2016;**58**(4):525-532

[64] Goodrich MA, Schultz AC. Human–robot interaction: A survey. Foundations and Trends in Human–Computer Interaction. 2008;**1**(3):203-275

Chapter 10

Perspective Chapter: European Robotics League – Benchmarking through Smart City Robot Competitions

Matthew Studley, Sarah Carter, Francisco J. Perez-Grau, Antidio Viguria Jiménez, Gabriele Ferri, Fausto Ferreira, Deebul Nair, Sven Schneider, Paul G. Plöger, Pedro U. Lima, Meysam Basiri, Gerhard K. Kraetzschmar, Daniele Nardi, Lun Wang, Emanuele Antonioni, Vincenzo Suriani and Luca Iocchi

Abstract

The SciRoc project, started in 2018, is an EU-H2020 funded project supporting the European Robotics League (ERL) and builds on the success of the EU-FP7/H2020 projects RoCKIn, euRathlon, EuRoC and ROCKEU2. The ERL is a framework for robot competitions currently consisting of three challenges: ERL Consumer, ERL Professional and ERL Emergency. These three challenge scenarios are set up in urban environments and converge every two years under one major tournament: the ERL Smart Cities Challenge. Smart cities are a new urban innovation paradigm promoting the use of advanced technologies to improve citizens' quality of life. A key novelty of the SciRoc project is the ERL Smart Cities Challenge, which aims to show how robots will integrate into the cities of the future as physical agents. The SciRoc Project ran two such ERL Smart Cities Challenges, the first in Milton Keynes, UK (2019) and the second in Bologna, Italy (2021). In this chapter we evaluate the three challenges of the ERL, explain why the SciRoc project introduced a fourth challenge to bring robot benchmarking to Smart Cities and outline the process in conducting a Smart City event under the ERL umbrella. These innovations may pave the way for easier robotic benchmarking in the future.

Keywords: robots, benchmarking, smart cities

1. Introduction

Technology has an increasing impact on all our lives, and as the population of the world becomes more predominantly urban, it is in cities that most of us will feel this impact. Two technologies which seem destined to shape our experience are Smart Cities and Robotics. Of course, for robots to be useful we need to trust that they will do their job properly and reliably, and this requires that we should be able to test them against standards, and see their performance with our own eyes.

Resulting from a series of initiatives funded by the European Union's Framework 7 and Horizon 2020 programmes, the European Robotics League (ERL) has been organising robot competitions for robots throughout Europe. Competitions are based around benchmarking, and aim to give robot developers data to enable the comparison of different systems, algorithms and approaches to solving complex tasks [1]. The ERL originally consisted of three challenges, which were suitable for Industrial, Service, and Emergency Robots (the latter challenge required that robots would work together in air, land and sea to solve a simulated disaster similar to that experienced in Fukushima). We have renamed the Industrial challenge as 'Professional' and the Service challenge as 'Consumer' to recognise the fact these sectors are converging around human and robot co-working with a focus on human-robot interaction.

One factor which defines the ERL's competition events is that they are most often in locations that are hidden away from the public such as robotics labs, test facilities, or disused ports. This distancing from the public is not deliberate and indeed not desired. Most often the competition events take place in locations which are determined by the need for accessible and safe indoor and outdoor environments that meet the needs of the benchmarking tasks at hand.

Surveys such as the Eurobarometer [2] have shown us that public attitudes towards robots are mixed, viewing them often with suspicion. It is unclear whether these attitudes are based on robotic fact or fiction; most people may not have had a direct experience of a robot. Since we wish to engage the public to raise issues around acceptance and desirability, and we wish to empower the people of Europe to direct robotic development, it seems clear that we should move our demonstrations of robots into venues that are easily accessible and where future use cases might involve people and robots working alongside each other.

It was for these reasons that the SciRoc project extended the European Robotics League to include the *Robotics in Smart Cities Challenge*. This allows teams to benchmark their robots and test their skills on a variety of tasks that one might find in a Smart City environment. Furthermore, teams need to interact with the Smart City data infrastructure while performing these tasks in order to demonstrate the value of the city host's investment and its readiness for future technology.

2. ERL processes for benchmarking and competition

We follow the process of designing competitions that has been outlined in the RoCKIn project and visualised in **Figure 1**. One of the driving forces behind a competition is the formulation of a challenge. In the SciRoc context those challenges build the foundation of the three individual leagues:

ERL Professional Service Robots (ERL Professional): How can mobile manipulation robots support low-volume, high-variety manufacturing?

ERL Consumer Service Robots (ERL Consumer): How can mobile manipulation robots support (elderly) citizens in their domestic environment?

ERL Emergency Robots (ERL Emergency): How can multi-domain robotic systems support rescue teams in emergency situations?

In all of these leagues, we follow the same methodology as outlined in the RoCKIn project;

DOI: http://dx.doi.org/10.5772/intechopen.108583

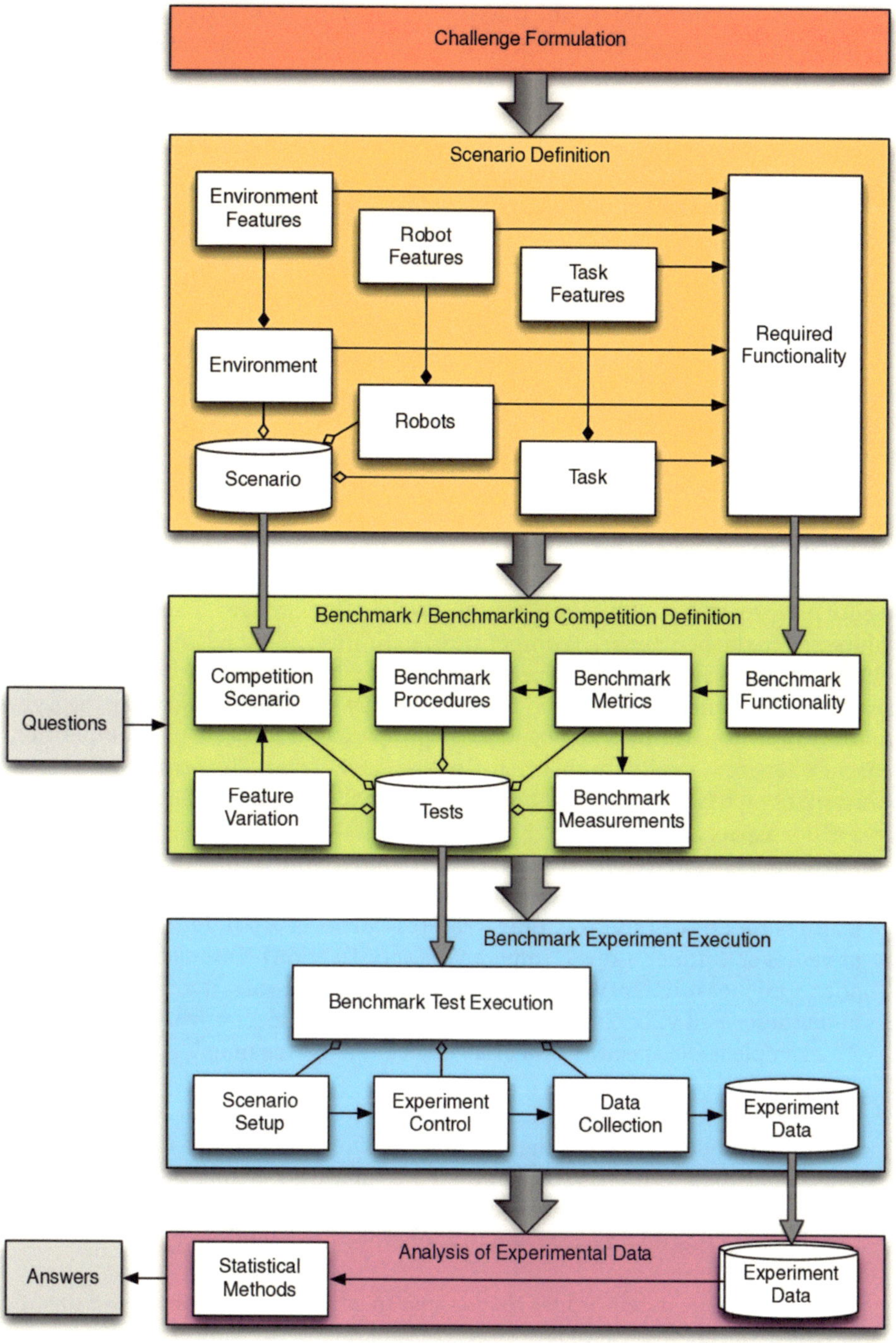

Figure 1.
Competition design process.

- Functionality benchmarks (FBMs), which evaluate the performance of robot modules dedicated to specific functionalities, in the context of experiments focused on such functionalities.

- Task benchmarks (TBMs), which assess the performance of integrated robot systems facing complex tasks that require the interaction of different functionalities.

SciRoc integrates all three leagues in a common scenario of a Smart City. This particular challenge revolves around the question of how autonomous robots can support citizens in various smart shopping tasks.

While the challenges frame the overall benchmarks and competitions they are too complex to be solved within a reasonable time frame like a single EU-funded project and too generic to be solved by individual researchers. Thus, as a next step, interesting scenarios are derived from the challenge. A scenario consists of the task to be solved, the environment in which this task will be executed and the robot that will solve the task. Each of those three categories is defined in terms of their features, that is, a specification of the possible or allowed variability.

From the perspective of a scientific experiment the features are free or fixed parameters. The definition of a scenario also gives rise to a list of (abstract) functionalities that a robot is required to have. Examples of such functionalities may include task planning, manipulation, grasping, visual object recognition, speech understanding or speech generation. It should be noted that the scenario definition purely describes an application domain, but, by design, is not meant to provide insights into a robot's performance in solving a particular task in that domain. This latter aspect is addressed in the benchmark (competition) definition phase where the goal is to define tests that evaluate a robot's performance. The first step is the definition of questions or hypotheses that should be addressed by the benchmark. Only the right questions allow constraint of the overall design space of a scenario to a competition scenario where few, explicitly chosen features can be varied. The feature variation characterises each benchmark test by a choice or instantiation of all relevant features. Three major aspects govern those choices:

- The specification aspect determines how the feature is described. For some features concrete properties and attributes can be asserted, for instance, in terms of values or enumerations ("imperative" specification). One example is the size of a room measured by width, length and height. In contrast, other features are better described via exemplifying representatives ("declarative" specification). An example is the specification of an object as a "coffee mug".

- The temporal aspect determines at which point in time the feature is instantiated. The choices vary from the scenario specification (e.g. the types of areas), over the testbed setup (e.g. the concrete size of areas), before the benchmark experiment execution (e.g. the location of manipulable objects) to during the benchmark experiment execution (e.g. outdoor light conditions).

- The control aspect determines the manner in which the parameters are specified or fixed. The choices vary from concrete values (e.g. runtime of an experiment) over ranges (e.g. allowed minimum and maximum size of areas) to unspecified conditions (e.g. outdoor weather).

Given different feature variations their impact on the robot's performance can be measured and compared via well-defined metrics. Similarly, benchmarks for the functionalities are defined. The steps to conduct the benchmark are described in the benchmark procedures. One artefact of particular interest that is compiled during this phase is the rulebook which communicates and clarifies the taken decisions and procedures to various stakeholders.

The next phase is the benchmark experiment execution where the tests are instantiated and robots solve the tasks in the real-world environment. The execution consists of the three activities (i) scenario setup, (ii) experiment control; and (iii) data collection. In the final phase, the resulting experimental data, also from multiple instantiations of benchmark experiments, is analysed to provide answers to the questions underlying the benchmark definition.

We now examine how the three European Robotics Leagues implement the same underlying methodology, and the ways in which this was expanded upon to create the Smart City Challenge.

3. ERL consumer robots

The European Robotics League Consumer Service Robots (ERL Consumer) is a research competition that aims at bringing together the benefits of scientific benchmarking with the attraction of scientific competitions in the realm of consumer service robotics. The objectives are to foster research in consumer service robotics for home applications and to raise public awareness of the current and future capabilities of such robot systems to meet societal challenges like healthy ageing and longer independent living.

Currently, ERL Consumer raises challenges in domestic environments that resemble similar challenges to be posed in smart city environments (e.g., shopping malls), such as interacting with humans, recognising and picking objects from shelves, bringing the objects to the human who requested it, and/or moving outside the home to populated areas, so as to prepare teams for the Smart City Tournaments. But ERL Consumer objectives can be adapted to other challenges, still within domestic environments or in other environments, e.g., hospitals.

All the ERL Consumer TBMs and FBMs stem from a User's Story developed during the pioneer EU FP7 RoCKIn project [3]:

> *Granny Annie is an elderly person, who lives in an ordinary apartment. Granny Annie is suffering from typical problems of ageing people: she has some mobility constraints. She tires fast. She needs to have some physical exercise, though. She needs to take her medicine regularly. She must drink enough. She must obey her diet. She needs to observe her blood pressure and blood sugar regularly. She needs to take care of her pets. She wants to have a vivid social life and welcome friends in her apartment occasionally, but regularly. Sometimes she has days not feeling so well and needs to stay in bed. She still enjoys intellectual challenges and reads books, solves puzzles, and socialises a lot with friends.*

For all these activities, ERL Consumer is looking into ways to support Granny Annie to live a full and independent life.

3.1 Environment description and testbeds

The ERL Consumer environment reflects an ordinary European apartment with all its environmental aspects, like walls, windows, doors, or blinds, as well as common household items, furniture and decoration. The apartment depicted in **Figure 2** serves as a guideline. It has

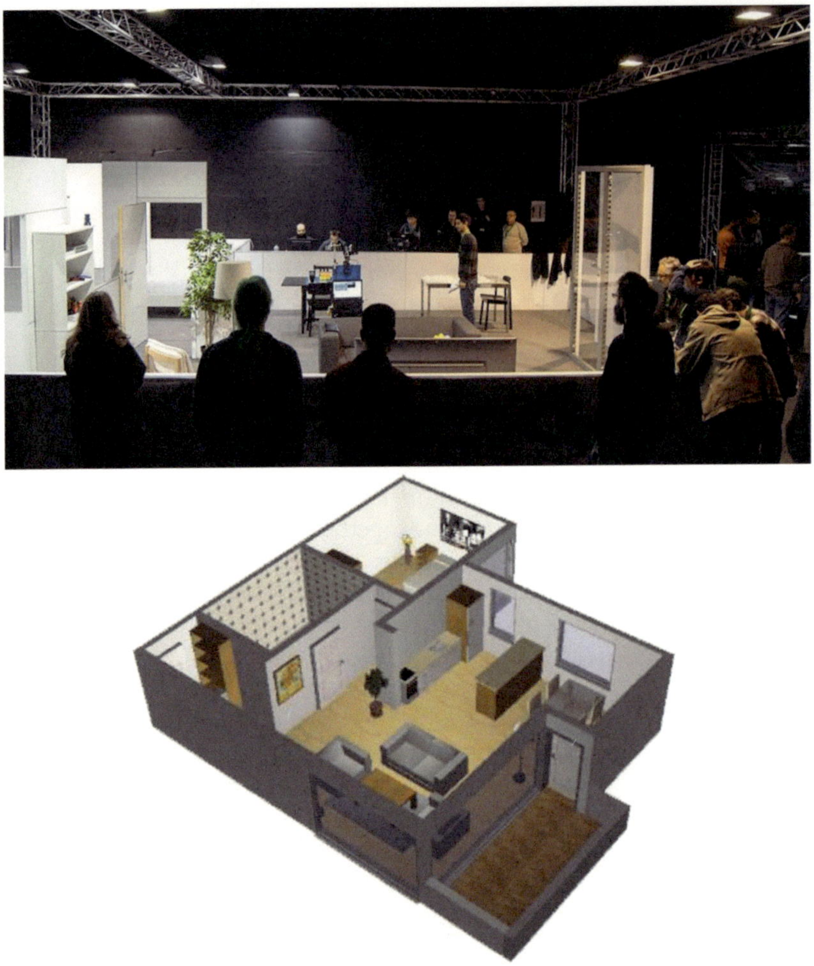

Figure 2.
ERL consumer testbed: On the left a real testbed during a RoCKIn@home tournament in 2015; on the right: The reference testbed as a simulated environment.

- Rooms (accessible to the robot): living room (with windows, couch, two armchairs, one coffee table, and one TV table), dining room (with one glass top dining table and two dining chairs), kitchen (with one kitchen table and two chairs, kitchen cabinet with multiple drawers and wash sink, two wall mounted kitchen shelves), inside hallway (with one coat rack), bedroom (with one window, a double bed, two side tables, two table lamps and one large wardrobe with mirror).

- Spatial areas (inaccessible to the robot): outside hallway, bathroom, patio.

- Well levelled floor, uniform all over the testbed, but including carpets.

- Wooden walls, most of them 50 cm high, but including one, behind the kitchen, 200 cm high.

DOI: http://dx.doi.org/10.5772/intechopen.108583

The furniture and available objects (e.g., glasses, forks, knifes, pillows, cups) were chosen to set up a long-term research program with challenges for robot navigation (e.g., mirror in the wardrobe, tables with metallic reflective legs) and perception (e.g., glass-top table, natural backgrounds).

In addition to furniture and decoration, the apartment is equipped with a computer network infrastructure:

- Server: a computer used to manage the network.
- Switch: an Ethernet switch used to connect all the networked devices.
- AP: An Access Point the mobile robot wirelessly connects to. Acts as a bridge between WLAN and LAN and provides access to a network of Ethernet cameras that provide perspectives of the home and of the outside hallway. Remote image acquisition is possible, and the camera parameters (frame rate, resolution, colour gains) can be changed over Ethernet.

Home automation embedded devices may be installed and are accessible within the apartment's WLAN, e.g.,

- the lamps in the bedroom (e.g., on the bed stand) are accessible and controllable via network;
- the shutters on the bedroom or living room window are accessible and controllable via network.

Eight testbeds, certified to follow the standard specifications of this environment, are currently part of the ERL testbed network:

1. ISRoboNet@Home Test Bed, at the Institute for Systems and Robotics in Instituto Superior Técnico, U. Lisboa, Portugal.
2. ECHORD++'s RIF @Peccioli, at The BioRobotics Institute, Scuola Superiore Sant'Anna, Peccioli, Italy.
3. Leon@Home Test Bed, at the Universidad de León, León, Spain.
4. BRL Anchor Personalised Assisted Living Studio, at the Bristol Robotics Laboratory, University of the West of England, Bristol, UK.
5. PAL Robotics Assisted House, at PAL Robotics S.L., Barcelona, Spain.
6. Heriot-Watt@Home Test Bed, at Heriot-Watt University, Edinburgh, UK.
7. IDEAAL Living Lab, at the OFFIS Institute for Information Technology, Oldenburg, Germany.
8. Cobot Maker Space Living Space, at the University of Nottingham, UK.

ERL Consumer tournaments take place in these testbeds. Some of the testbeds include a Motion Capture (MoCap) System, which enables tracking with high accuracy robots, people and object locations. MoCaps are mostly used in the FBMs.

3.2 Task benchmarks

Currently, ERL Consumer includes 4 Task Benchmarks:

- **Getting to know my home**: The robot must detect new changes in the environment, and update a semantic map of the apartment, within a limited time frame. The task is performed by the robot autonomously, though it may include moments of symbiotic interaction with a user in the apartment, e.g., to learn more about an object or a location. At the end of the knowledge acquisition phase, the robot must show the understanding of the new environment, namely by addressing changed objects and furniture locations, handling one of the changed objects between two furniture locations, one of them with its original location changed.

- **Welcoming visitors**: The robot needs to handle a set of known and unknown visitors, who arrive individually at the home entrance in random sequence. The robot must correctly recognise (using the networked camera over the outside hallway) the known visitors and interact with the unknown visitors to identify them and understand their purpose of visit. The robot must then perform a set of visitor-specific behaviours that could range from manipulating and delivering of objects to guiding and following the visitors.

- **Catering for Granny Annie's comfort**: This task aims at providing general purpose requests of Granny Annie inside the apartment. It focuses on the integration of different robot abilities such as human-robot interaction, navigation, and robot-object interaction. The robot is required to understand the actions requested by speech (such as finding, picking, and bringing an object to Granny Annie) and execute them accordingly.

- **Visiting my home**: This TBM focuses on safe navigation in dynamic environments, people perception, obstacle avoidance and tracking and following a human. In this task, the robot should visit a set of predefined rooms, to perceive and count the number of people in each room, while avoiding and/or interacting with different obstacles based on the nature of the obstacle. Furthermore, the robot must interact and follow a previously unknown person outside the arena through a small crowd and then guide that person back to the arena.

3.3 Functionality benchmarks

Currently, ERL Consumer includes 6 Functionality Benchmarks [4, 5]:

- **Object Perception**: Evaluates the ability of a robot to recognise and localise a wide range of objects. A set of objects, selected from the list of ERL Consumer items, is positioned, one at the time, on a table located directly in front of the robot. For each object presented, the robot must perform the following activities:

i) Object detection: perception of the presence of an object on the table and association between the perceived object and one of the object classes. ii) Object recognition: association between the perceived object and one of the object instances belonging to the selected class. iii) Object localization: estimation of the 3D pose of the perceived object with respect to the surface of the table (ground-truth provided by the MoCap).

- **Navigation**: Evaluates the ability of a robot to correctly, safely, and autonomously, navigate in an ordinary apartment, including: the navigation within furniture, walls, and doors, in a previously mapped area; avoiding collisions with different types of unknown obstacles, in unknown positions (not previously mapped); and navigating in the presence of people in the arena.

- **Speech Understanding**: Evaluates the ability of a robot to understand speech commands that a user gives to a robot in a consumer environment, including all the related issues, such as background noise caused by other ongoing activities and by the robot motion. A set of spoken sentences, recorded in different environments, is broadcasted through a speaker. The robot needs to interpret the commands and produce an output according to a defined representation.

- **People Perception**: Evaluates the ability of a robot to locate and recognise humans. Similarly to object perception, the robot must recognise a person who is standing inside a target area and to estimate the location of this person.

- **Person Following**: Evaluates the ability of a robot to effectively follow a human target around and through obstacles and a crowd of walking people. The benchmark requires that the robot accompanies a human target and always maintain a desired distance with this person.

- **Grasping and Manipulation**: Evaluates the ability of a robot to correctly grasp and manipulate objects. In particular, it assesses the object picking and placing capabilities of robots suitable for many consumer applications such as setting up a dining table in domestic environments.

4. ERL professional service robots

ERL Professional Service Robots (ERL-PSR) League is focused on the major challenges addressed by H2020: industrial robots addressing the flexible factories of the future and modern automation issues.

Greater automation in broader application domains than today is essential to ensure European industry remains competitive, production processes are flexible to custom demands and factories can operate safely in harsh or dangerous environments. In the ERL-PSR competition, robots will assist in filling stocks in a department store. The task includes locating, picking, transporting and placing them in the proper shelfs. The combination of human versatility and reliability of mobile robots will optimise the entire process. The ERL-PSR competition is looking to make these innovative and flexible manufacturing systems, such as that required by the smart factory, a reality. This is the inspiration behind the challenge and the following scenario description.

In this version of the ERL-PSR we focus on the problem of picking products from a shelf and placing them in the shopping basket. The domain is a common department store, where the robot helps to arrange the inventory of the department store on the shelves. The main functionality tested in this episode is mobile manipulation using an autonomous robot. In recent years, mobile manipulation has become a problem of interest among researchers due to the variety and complexity of challenges and robot capabilities that are involved. For instance, in a grocery store setting, there are so many complexities that a robotic hand or gripper should take into account, such as handling objects of different weight, sizes, shapes, texture and compliance. Picking up a bag of pet food is very different from picking up bananas or cucumbers. Different robotic competitions [6–11] have picking and packing problem, aiming at 1) measuring the performance of these complex systems, 2) defining metrics and assessing benchmarking criteria, 3) fostering research and development of new approaches and technologies, and 4) creating awareness and drawing more attention from the general public to robotics.

4.1 Environment description and testbeds

The ERL-PSR testbed consists of different elements which include the arena, networked devices, department store shelves and the products in the store. The robot can communicate with the department store inventory management system and to other networked devices. The robot receives tasks to perform from the inventory management system.

The following set of scenario specifications must be met by the ERL-PSR environment.

The environment can consists of various numbers of spatial areas:

1. rows of shelves

2. rows of workstations

Figure 3-a shows an example of these areas in the ERL-PSR environment. The spatial areas extend beyond the space occupied by the respective workstations or objects and include the surrounding area as well.

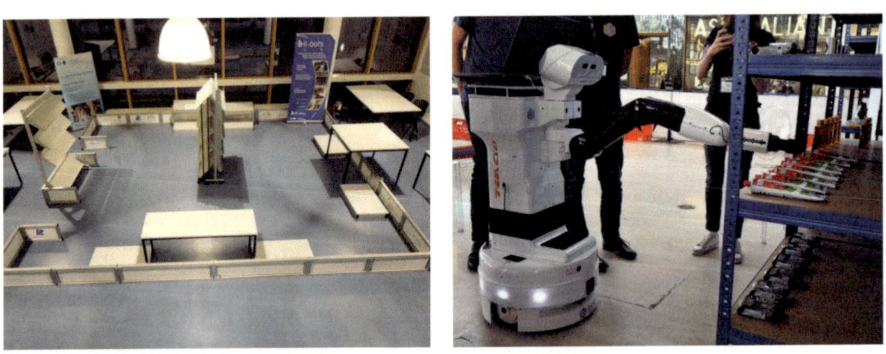

(a) (b)

Figure 3.
ERL PSR different testbeds. a) ERL PSR testbed, Sankt Augustin, Germany. b) Testbed with objects and shelves.

All spatial areas are located on the same level, except where specified otherwise. There are no stairs in the environment. The environment is a replica of department store with aisles of shelves containing different objects. The environment has a boundary. The precise dimensions and the arrangement of the spatial areas are not predefined, but estimated sizes are given. The estimated sizes of the spatial areas are as follows: workstations 2 m × 2 m and shelves 5 m × 0, 5 m. The bounding box of the environment has a minimum area of $16m^2$ and a maximum area of $100m^2$. More space is used, when areas and workstations are doubled for teams working in parallel. Workstations are used as storage areas for objects. They may be accessible from different locations, i.e. it might be possible to reach a workstation from two or more sides. Additionally, there are workstations of different heights present in the environment, ranging from 0 cm up to 15 cm. If a workstation has a height of 0 cm, a tape will mark the area (see **Figure 3**). The tape will be taped on the floor and is blue/white striped. This tape may be crossed by the robot and does not count as a collision.

4.1.1 Objects in the environment

The objects to be manipulated will be selected by the organiser for each tournament. The following lists describe the different categories of objects that can be used by the organiser during a tournament depending upon the scenario being chosen. The different categories of objects are (**Figure 4**):

1. RoboCup objects

2. Ocado objects

3. Chocolate objects

4.1.2 Robots and teams

A competing team can use single or multiple robots. The robots are not required to be certified for industrial usage. The robot should have at least the following capabilities:

- autonomous navigation.

- grasping capability wireless communication capability(802.11 version 5Ghz).

- safe for public usage.

RoboCup Objects | Chocolate Objects | Ocado Objects

Figure 4.
Objects in environment.

The different subsystems of the robot should work in the environment and manipulate the objects specified in the rulebook. The teams can use any sensor available in the market provided they are safe to use with humans and have the corresponding certification. The teams are not allowed to modify the arena which is finalised by the organiser. The teams are allowed to use any internal communication protocol. The robot shall pass through a safety test before it is admitted in the arena.

4.2 Task benchmarks

4.2.1 Fill a box with parts

This task is one of the primary task in a department store where customers fill their basket with products from shelves. The robots have to deal with flexible task specifications, especially concerning information about object constellations in source and target locations, and task constraints such as limits on the number of objects allowed to be carried simultaneously, etc. The robot has to pick up several parts from different source locations and deliver them to several destination locations.

4.3 Functionality benchmarks

The task benchmark was subdivided into multiple Functionality Benchmarks. The list of benchmarks executed are:

- **Object Detection**: this functionality benchmark evaluates robot capabilities of locating an object at a given location. One of the common tasks for service and industrial robots is to locate the object which can possibly be placed at a particular location. In addition to that, several secondary objects and decoys may be present at the location too. The robot is required to find particular objects among a set of objects and decoys. The target object is either included or not depending on the variation for every trial.

- **Manipulation Pick**: this functionality benchmark evaluates the grasping capability of the robot. The robot has to identify the object in front of it and then attempt to grasp it and pick it up. Once the object has been picked up the robot has to notify.

- **Manipulation Place**: this functionality benchmark evaluates the placing capability of the robot. Based on the task different objects have to be placed in different orientation. The robot based on the task has to identify how to place the object and then notify about the status of the placement.

- **Exploration**: this benchmark evaluates the robot exploration and navigation capability. The test benchmarks the navigation capability of the robot to simultaneously explore the environment and perceive the environment for a particular object present in the environment.

5. ERL emergency robots

ERL Emergency tournaments challenge multi-domain teams of marine, land and aerial robots with search and rescue tasks, inspired by a disaster response scenario [6].

Following the experience of the euRathlon 2015 Grand Challenge [7], the ERL Emergency competition started in 2016 and culminated in the ERL 2017 Major Tournament [8] held at the Tor del Sale power plant site of Piombino, on the coast of Tuscany (Italy). The inspiration came from the 2011 Fukushima accident [9]. Land, marine and aerial robots cooperated in a disaster scenario organised at a real power plant, and were required to survey the accident area, to identify and help missing workers (mannequins) and to intervene to stem a leak by closing valves both inside the building containing the machine room of the plant and underwater.

Similar concepts were followed in the SciRoc project, and in 2018 and 2019, ERL Emergency moved to a model of local tournaments with two-domain competitions: land + underwater and land + aerial robots competitions. The Centre for Maritime Research and Experimentation (CMRE) organised two editions of the land + underwater robot competition at its premises in La Spezia (Italy) in 2018 and 2019. The land + aerial competition was hosted by the Advanced Centre for Aerospace Technologies (CATEC) at Seville (Spain) in 2019. These events were a preparation for the Smart City events, where only the tasks of the ERL Emergency involving the aerial robots were present.

Over the years, ERL Emergency has proposed team tasks which required advanced perception skills (to locate numbers positioned underwater, underwater pipes, or persons for first-aid kit deliveries using aerial robots), intertwined with autonomy and cooperation capabilities in realistic scenarios. Multi-domain cooperation has been specifically searched and pushed. Robots are required to be able to accomplish adaptive missions, for instance executing different actions on the basis of different sensed conditions. The increased attention to autonomy and cooperation has also been accompanied by an increasingly metrological attention to benchmarking the robot performance. This has been achieved by separating the Task Benchmarks from the Functionality Benchmarks, following the general trend of the ERL evaluation framework [10].

5.1 Land+underwater robots events held at CMRE sea water basin

In the SciRoc project, CMRE conducted two ERL Emergency local tournaments, in 2018 and 2019, challenging multi-domain teams composed of an underwater and a land robot with a scenario of emergency response to a simulated explosion in a harbour. The areas of operations included a building with the surrounding outdoor space for land robots, and the CMRE sea water basin for underwater robots (see **Figure 5**).

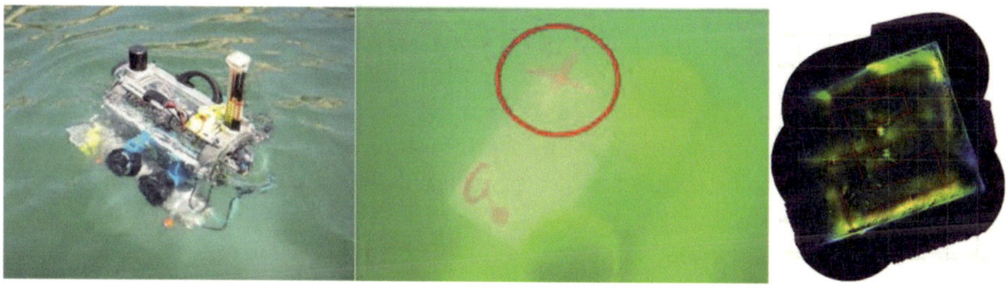

Figure 5.
(left) the Feelhippo AUV from the UNIFI robotics team in action. (Centre) the yellow pipe structure with the manipulation console used in the ERL emergency competitions. (right) a multi-beam mosaic by team Tomkyle of the competition area. The different objects of potential interest (buoys, pipeline assembly structures, the gate are visible).

The marine robots were challenged with the realistic conditions typical of at-sea operations (water salinity, changing light conditions, waves and tides).

The scenario represented an accident of a yacht in a harbour. The vessel clashed on the dock causing an explosion. The accident affected the area around the harbour, both outdoor and a building. The land robots (unmanned ground vehicles – UGVs) were tasked to survey the outdoor area, localise pipes and find a missing worker (represented by a mannequin). The robots were requested to deliver a first-aid kit in proximity to the mannequin. Successively, inside the building, the robots were required to localise and close a valve and to find a canister. The specific valve to be closed was communicated to the UGV by the underwater robot of the same team. UGVs needed to transport the canister from inside the building to a simulated outdoor fire location. Autonomous underwater vehicles (AUVs) had to pass through a validation gate - composed of two submerged buoys, survey an area to detect a missing person underwater (a realistic mannequin) and localise an emitting pinger (see **Figure 5**). They also were required to inspect a pipeline structure and localise a damage (represented by a marker), at the same time reporting its shape and size. Underwater robots had to be autonomous, which means that all navigation and perception tasks were needed to be accomplished autonomously as well. Buoys of different colours were to be identified, localised and their colour recognised. The robots had to perform a different action, depending on the buoy colour: as an example, turning in a clockwise circle around the buoy or stopping for 30 seconds increasing the depth. The objective was to push teams to integrate perception with adaptive and reactive mission planning in a realistic scenario such as presented in the CMRE water basin. Here the changing and real conditions, such as the limited visibility underwater, created severe difficulties for object recognition by robots, even in the case the buoys were bright orange or red in colour.

Our experience suggests that one of the important aspects in real-world competitions is the possibility to guarantee teams sufficient time for practicing and tuning their systems to the changing and complex environmental conditions. This is especially true for marine robots, since teams may have difficulties to have access to sea waters for testing before the event. To allow teams to test, a practice arena was prepared.

As a way to increase the level of the challenge, the size of the objects to be detected was reduced with respect to previous competitions. Some of the tasks and functionalities were common to the ERL Emergency Local Tournament in Seville (land + aerial robots), such as the object recognition or map building functionalities. This allowed for a similar comparison of results in both competitions for a given Functionality Benchmark (FBM) (e.g. object detection). The same kind of objects of potential interest were used so the Object Detection FBM could be evaluated in both competitions. Some of the tasks were required to be completed with robots operating in a fully autonomous way, while others could be accomplished remotely controlled (e.g. with the assistance of an operator). However, no manual perception tasks were allowed. All object detections had to be either autonomous (real-time by the robot) or automatic (offline by a computer).

In both events, a mix of well-experienced teams and new teams attended the competitions. In 2018, five teams participated (three for marine domain, one for land and one deploying robots for both domains), while in 2019 [11], the number of teams increased with the participation of seven teams (four for marine domain, two for land domain and one with both segments).

Results highlight the improvement of team performance over the years, especially for teams which succeeded in participating in multiple editions of our events. Our policy to increase the task difficulty year after year led in fact the teams to increase their technical and, above all, management skills, thanks to the gained experience from their unavoidable errors and from the interactions with other entries. For these reasons, it is important to guarantee a continuity in the organisation of the competitions, especially when robots have to handle real-world scenarios in non-controllable conditions, which dramatically increase the task difficulties.

5.2 Land+aerial robots events held at CATEC

CATEC organised an ERL Emergency local tournament in February 2019, targeting aerial and land robots working in an outdoor/indoor environment [12]. The scenario was an earthquake in an industrial area near a factory building, where a robotic team composed of land (UGV) and air robots (UAV) had to intervene. The priorities are to discover and assist missing people, and determine if the building has suffered any serious damage. The robots have to search for missing workers (one outdoors and one indoors), find them as soon as possible and deploy an emergency kit to both of them. Besides, the robots must check the state of the building after the earthquake, for which a detailed map is required to assess the safety of the area (see **Figure 6**).

As mentioned before, some of the functionalities were shared with the land + underwater robot competitions. In this case, the same kind of markers were used as objects to be recognised, so the Object Detection FBM could be properly evaluated in both competitions. From the starting points, land and aerial robots must inspect the area, autonomously detecting and avoiding obstacles while traversing the competition arena. Then, they needed to find a suitable entrance that could be used to enter the building, using similar markers as the ones used for object recognition. While accessing the building, robot navigation must deal with transitioning from an outdoor to an indoor environment. Once inside, robots had to build a detailed map of the scenario, either 2D or 3D according to their sensor suite. During the whole mission execution, as soon as a missing worker was found (represented by a mannequin), the robot provided a first-aid kit as soon as possible.

The setup for the competition was an area of approximately 200 m x 30 m free of obstacles, in an area where operating UAVs in Visual Line Of Sight (VLOS) is not forbidden according to regulations in Spain. The building was represented with a 20 m x 20 m marquee where the aerial and ground robots need to access. Static obstacles (e.g.

Figure 6.
(left) the UAV and UGV from team LARICS in action, reaching the outdoor missing worker. (Centre) the rover UGV and the UAV from team raptors looking for the indoor missing worker. (right) top view of a 3D map of the competition area by team LARICS. The trajectory followed by the robot is shown in blue.

stones, holes, vegetation...) and dynamic obstacles (e.g. birds...) were expected in the outdoor area, as well as loss of Wi-Fi signal. As with any outdoor competition, there was also the possibility of rain, wind, dust and muddy areas, but weather was generally fine during that week. A practice area was also set up to provide teams enough space for testing their systems before deploying them in the competition arena.

5.3 Task benchmarks

TBMs in ERL Emergency require the cooperation of robots in different domains. This often requires teams to work together.

5.3.1 Yacht accident in the harbour (land+sea)

This two-domain task benchmark is focused on acquiring knowledge about the environment and its explicit representation; and to cooperate between domains to search for the missing workers and give them assistance. The ground and underwater robots are required to understand the changes in the environment and interact with it either through cooperation between them (autonomous robot-robot) or their operators (human-robot interaction) or with a mixed approach. A minimum of one land robot and one underwater robot is required to participate in this task.

The motivating scenario is as follows;

> *An accident occurs in the harbour when a yacht arriving at the pier damages a gas pipeline which leaks and causes an explosion. This also affects the building containing the pipeline section on land and people that were in the docks area are dispersed. The emergency response team arrives soon but members must maintain a safe distance from the fire. For this reason, the use of robotic vehicles is essential. A robotics team composed of land (UGV) and underwater robots (AUV) is ready to intervene.*

Three missions can be undertaken.

- Mission-A: Search for missing workers. Locate and help missing workers, removing rubble trapping them.
- Mission-B: Reconnaissance and environmental survey. Provide situational information to the emergency team, exploring the damaged building and underwater parts of the damaged pier.
- Mission-C: Pipe inspection and stopping the fire. The correct valves must be turned to stop a simulated fire, both underwater and in the damaged building.

5.3.2 Emergency in a building (land+air)

This two-domain task benchmark is focused on acquiring knowledge about the environment and its explicit representation; and to cooperate between domains to search for the missing workers and give them assistance. The ground and aerial robots are required to understand the changes in the environment and interact with it either through cooperation between them (autonomous robot-robot) or their operators (human-robot interaction) or with a mixed approach. A minimum of one land robot and one aerial robot is required to participate in this task.

The motivating scenario is as follows...

> *An earthquake has occurred in an industrial area near a factory building, and two workers are missing (one inside, and one outside). The emergency response team arrives soon but members of the response team must maintain a safe distance from the building. For this reason, the use of robotic vehicles is essential. A robotic team composed of land (UGV) and air robots (UAV) must discover any missing people and whether the building has suffered any serious damage. Robots have to find wrokers as soon as possible and deploy an emergency kit to each. Robots must check the state of the building after the earthquake and create a detailed map for the emergency team.*

This task benchmark comprises two missions' goals:

- Mission-A: Delivery of emergency kits to missing workers. The two workers must be found as quickly as possible, and first aid kits deployed.
- Mission-B: Mapping for safety assessment of the building. Both land and air robots must collaborate in creating a 2D or 3D map of the space inside.

5.4 Functionality benchmarks

- **2D Mapping Functionality (Land):** This measures a land robot's ability to explore a 2D area while visiting a number of waypoints, and is scored according to map coverage and accuracy of the waypoint locations.
- **Mapping Functionality (Air):** This measures an aerial robot's ability to explore the competition area while visiting a number of waypoints, and is scored according to map coverage and accuracy of the waypoint locations.
- **Vertical Wall Mapping Functionality (Sea):** This assesses the capabilities of marine robots in extracting information about a specific wall of the damaged pier. The identity of the wall to be explored will be communicated to the underwater robot by the ground robot. After the exploration, teams must provide a 2D or 3D map of the designated wall along with several measurements calculated from the map.
- **Object Recognition Functionality (Land):** This assesses the capabilities of ground robots to recognise objects that might be found in an outdoor and indoor disaster response environment. The benchmark requires that robots detect Objects of Potential Interest (OPIs) and identify the type of each object found, and is scored according to the precision and accuracy of identifying and locating the objects.
- **Object Recognition Functionality (Sea):** This assesses the capabilities of underwater robots in extracting information about observed objects. The objects to be recognised in this FBM are the orange buoys that act as obstacles. Each obstacle buoy is identified by a black number, from 1 to 4, and scores depend on the number of buoys correctly detected, identified, and accurately located.
- **Object Recognition Functionality (Air):** This assesses the capabilities of aerial robots to recognise objects that might be found in an outdoor and indoor

disaster response environment. The benchmark requires that robots detect Objects of Potential Interest (OPIs) and identify the type of each object found, and is scored according to the precision and accuracy of identifying and locating the objects.

6. Smart cities competitions

The label 'Smart Cities' is used with increasing frequency with a main focus on the use of engineering approaches within the built environment to improve citizens' quality of life. These visions present the city as an entity, which gathers and processes data to decision makers and infrastructure, and which offers data services to enable and improve services offered by other stakeholders such as private companies. Since the city senses its environment, processes this data, and takes action through effectors like traffic signals, the Smart City resembles a Robot. We might expect to find that Smart Cities and Robots gain mutual benefits from their integration, and also find some common problems which both must overcome.

The world's population is becoming more urban, so it is likely that most human-robot interactions will take place in Smart Cities. Therefore, it seems that any ethical challenges that arise from the mutualistic interactions of Robots and Smart Cities will have a significant impact for these future citizens. How can robots act best to sustainably promote our happiness and prosperity?

When we first stated our intention to bring a scientific robot competition to the heart of a major public space, our ambition was to address this question by stimulating public discussions based on robotic facts about the roles which robots might play in the future in their smart city. Our motivation was to demonstrate that robots and smart cities are natural partners which add mutual value, and in doing so to showcase the state of the art in European Robotics to benefit the teams, sponsors, and wider community. We were able to achieve our aims through the successful delivery of two ERL Smart Cities Challenges to date. The first in Milton Keynes, UK in September 2019 and the second in Bologna, Italy in September 2021.

Our intention was to provide for the sustainable future of the ERL beyond the H2020 funded SciRoc project, and as part of this our first competition was led by a project partner working with the city, and with some funding from the project budget, while the second was led by a third party and entirely self-funded. This demonstrates that this mode of public engagement with benchmarking through competition can be repeated using this model in the future.

In the following sections we explain how and why the TBMs and FBMs of the ERl were adapted and extended to create the Smart City Episodes. We also explain how the critical elements which need to be considered for a city host looking to implement a future Smart City event, or similar urban robotic benchmarking challenge.

6.1 Milton Keynes, 2019

The first Smart City event took place at the Milton Keynes Centre: MK shopping mall over five days, providing teams the opportunity to compete in five different episode scenarios. This public space attracts on average over half a million people each week, and many of these people paused to watch the robots as they competed. The event's sponsors included PAL Robotics and Ocado, and attracted 11 international robotics teams from different Universities throughout Europe.

Allied events included a Symposium to debate the risks and opportunities associated with the emergence of a "hybrid society", and a Workshop on the Evaluation and Benchmarking of Human-Centered AI Systems. This public engagement helped foster public debate and equip citizens to be engaged active stakeholders in their city's future.

6.2 Bologna, 2021

The second Smart City event took place at Palazzo Re Enzo, in the heart of Bologna. In contrast to the first Smart City event in Milton Keynes, this competition was organised by the host city, which provided all the resources to run the event, assisted by technical and scientific support from the SciRoc Consortium to set up and run the five episodes.

The implementation of the competition had to face significant uncertainty caused by the Covid-19 pandemic. Challenges arising through unforeseen travel restrictions were addressed with innovative technical solutions in terms of the arrangements of the competition, to allow teams and robots to compete both locally and remotely.

Such measures included the introduction of a simulation environment for a service robots scenario; remote participation on a physical robot made available on site; introduction of a novel challenge related to sign language that had outstanding success within the deaf community; synergy with the H2020 Eurobench project aiming at implementing benchmarks for robots; design of a new challenge for emergency robots; and finally distributed execution of the competition with both on site and remote participants performing the tests at their own labs.

6.3 Hosting a Smart City event

In this section we briefly describe the process of hosting a SciRoc Smart City competition in the hope that this assists readers interested in running a similar benchmarking event, or hosting a future Smart City Event.

6.3.1 The selection process

Smart City hosts are determined through an open call process. Interested candidates can register their interest in response to the call and are required to submit a proposal for the event, and complete an interview with the selection committee. The selection committee, comprising members selected to remove conflicts of interest, determined the host through a final vote.

6.3.2 Competition design

The competition is divided into a series of robotic challenges, referred to as episodes. This represents an innovation of the structure of the competition, based on Functionality Benchmarks (FBMs) and Task Benchmarks (TBMs). Episodes are intended to be an intermediate challenge between the local tournaments' FBMs and the TBMs. Even though the methodology and the approach to the competition are de facto unchanged, the proposed format based on episodes is designed to make a step forward towards their application in a realistic scenario, to allow for a better communication towards the general public and to lower the entry level for the competing teams.

An episode places a given functionality, tested during a specific FBM in one of the other ERL challenges, into a more social and operational context, while limiting the amount of effort needed for their development.

Episodes are organised into categories depending on the task to achieve and the type of robots involved. We group them into three categories:

1. **HRI & Mobility**: all episodes involving robots able to show social behaviours, such as verbally interacting with (human) customers or navigating respecting proxemics, in line with the current ERL-Consumer. These are meant for any robot (wheeled, or legged) with navigational and verbal communication.

2. **Manipulation**: this category includes all episodes requiring robots to achieve manipulation tasks, applying ERL-Professional services to the smart city context. Episodes here are meant for any robot able to navigate and equipped with arms and effectors for the manipulation of objects.

3. **Emergency**: the last category comprehends tasks and challenges addressed by small VTOL aerial robots, along the lines of the ERL-Emergency. Any VTOL aerial platform able to carry and deliver items in specific locations, navigate in outdoor and indoor spaces, detect and avoid obstacles can take part in episodes of this category.

6.3.3 Funding

The event is primarily funded through sponsorship funding, although it is expected that the host city will make their own contribution. For example, at the 2021 event in Bologna, the Municipality allowed the use of the competition venue Palazzo Re Enzo at no cost. This contribution was worth 65,000 EURO in kind. The University of Bologna also contributed 36,000 EURO in cash sponsorship towards the event.

6.3.4 Logistics

The host city is responsible for organising the venue for the event. When selecting the venue, organisers should consider that the location must be able to accommodate a number of factors, including: infrastructure for a minimum of five episodes, accessibility for teams to move robots around the venue, the potential use of drones within the venue, secure storage areas for robots and equipment, to name a few.

In addition to the competition venue, the host city should consider the following requirements for the functioning of the event: robot transportation, accommodation, financial support for teams, co-located events, the organising team and roles required within and the data hub.

6.3.5 The data hub

Central to emerging Smart Cities are online platforms for data sharing and reuse, which are normally called Data Hubs. Such Data Hubs have two purposes. They provide static datasets, such as demographic data which changes slowly, and dynamic live data gathered from sensors deployed within the city. The latter could include data about traffic flows, the environment, and the movement of people and robots, etc.

The MK Data Hub is a state of the art computational infrastructure, which supports the acquisition and management of city data. Specifically, it provides both a catalogue of several hundred data sources, as well as a development environment to facilitate the creation of data-intensive applications. The MK Data Hub played a key role in the 2019 SciRoc event in Milton Keynes by allowing us to simulate the diversity of systems with which robots were asked to interact throughout the competition.

Going forward, this approach could contribute a valuable additional output of robot competitions. First, by enabling the dynamic simulation of the environment in which the robots operate, the MK Data Hub introduced a contextual element in the benchmarks.

Second, a Data Hub environment provides the opportunity to record a large amount of heterogeneous information about the robots' behaviour. Such information can be reused (a) to enable a deeper analysis of the shortcomings of robots, (b) to compare the performance of the same robot in different trials, and in different competitions among the years, as well as (c) for archival purposes. Ultimately, recording robot messages and analysing their behaviour may constitute the first step towards benchmarking elements such as self-awareness and deliberation capabilities.

6.3.6 Teams

Teams are recruited through an open call process, which is disseminated via the SciRoc website and social platforms, mailing lists, and by the SciRoc Consortium within academic institutions and the Robotics community. All previous year's teams are also contacted directly to notify them of the open call.

6.4 Scientific contributions

SciRoc had the ambition to provide a testbed for experimental evaluation of Human-Robot Interaction (HRI) approaches. In fact, competitions offer a unique opportunity to create interaction scenarios for evaluating the performances of different approaches to human-robot interaction. In this respect, the work carried out in SciRoc had an impact on the HRI research community: the results of the experimental evaluation carried out in the SciRoc episodes, specifically designed for this purpose, have been published in the top venues for HRI research. Specifically, the Elevator episode of SciRoc 1, where the robot had to interact with users in taking an elevator, was accompanied by the creation of an ad hoc questionnaire [13] and the data collected during the competition have been used to demonstrate the potential for carrying out experimental studies within robotic competitions [14, 15]. Moreover, the Sign Language episode of SciRoc 2, where the robot was supposed to interpret and produce phrases of the Italian sign language, was accomplished with an unprecedented collaboration and impact with the deaf people community [16].

6.5 The continuation of Smart City events in the future

Results from teams are encouraging, and successful attempts at challenges of annually increasing difficulty suggest that teams have grown in capability through their repeated engagement with the benchmarking process. We have underlined the importance of continuing the organisation of such events for supporting the growth of teams and research groups, providing them with an annual real-world ground where to test their systems. This is especially true after the lockdown caused by the

COVID19 pandemic, which increased the difficulties for teams to access real-world training areas (in particular marine sites).

Although the SciRoc project will finish in 2022, we are working to continue the ERL and the Smart City Events after the project's close. The euRobotics international non-profit association will adopt the Smart City Events and continue to run them through an open call process, as with our previous Bologna event. It is intended that the design of the SciRoc events are responsive to the needs of the city stakeholders and, to some extent, the events are designed in order to support the communications and ambitions of the host city. We are therefore confident that city partners will continue to respond to our open call, to host Smart City events in the future.

The SciRoc Smart City Events saw collaboration with the H2020 funded EUROBENCH project to extend its work on benchmarking humanoids, and the H2020 METRICS project has been adopted within the ERL brand, following ERL benchmarking methodology. We hope that this will be repeated in the future, and anticipate new ERLs. Future Smart City competitions will provide a venue for more European Robotics projects to showcase their outputs to the public.

7. Conclusions

In this chapter we presented the SciRoc project which continued the work of the ERL and extended it to Smart City Events. We showed how the common TBM & FBM approach of the ERL was adapted and extended to enable these events to increase the value of the benchmarking endeavour through a massive increase in public engagement. The Smart City events delivered value to multiple stakeholder groups;

- Municipalities showed their city to be a future-ready venue to attract technological innovation and investment, and demonstrated the value of their smart city infrastructure

- Citizens and End Users saw robots performing relatable and believable tasks in accessible environments, which enabled fact-based discussions of fears and ambitions

- Host Institutions demonstrated their value in the political and economic ecosystems of their region, and showcased their capabilities to a wide and diverse audience

- Sponsors promoted their brand regionally and nationally. The value they gained beyond this depended on the nature of their business, for example, by building their network of top-quality robotics researchers

- Teams tested their robots, demonstrated their skills, and had fun while they did it

- Researchers extended the reach of their public engagement, developed and tested new benchmarking techniques and ways of working, and strengthened their networks

Bringing Robotics Benchmarking into the public arena in this way is important, we feel. For robots to be useful, they must be trusted, and for this trust to grow, real robots must be seen and their capabilities assessed.

Due to the restrictions placed upon us by the Covid-19 pandemic and the associated restrictions in travel and public assembly, we had to find new ways of working which would allow the teams flexibility in the ways they interacted with the Smart City Episodes. These methods may provide a basis for future development of benchmarking, and point the way towards a mixture of physical and virtual assessment.

As the project comes to an end, we are happy that we have surpassed our objectives, overcome unforeseen challenges, and that we have prepared the way for the continuation of the ERL through the support of the European Scientific Community and other stakeholders in the years to come. We hope that the trend to harmonise benchmarking under the ERL umbrella continues, and that the Smart City Events provide an arena for more European Projects to meet the public and therefore advance the successful and safe use of robotics within Europe and beyond.

Author details

Matthew Studley[1]*, Sarah Carter[1], Francisco J. Perez-Grau[2], Antidio Viguria Jiménez[2], Gabriele Ferri[3], Fausto Ferreira[4], Deebul Nair[5], Sven Schneider[5], Paul G. Plöger[5], Pedro U. Lima[6], Meysam Basiri[6], Gerhard K. Kraetzschmar[5], Daniele Nardi[7], Lun Wang[7], Emanuele Antonioni[7], Vincenzo Suriani[7] and Luca Iocchi[7]

1 University of the West of England, Bristol, UK

2 Advanced Center for Aerospace Technologies (CATEC), Seville, Spain

3 Centre for Maritime Research and Experimentation (CMRE), La Spezia, Italy

4 Faculty of Electrical Engineering and Computing, University of Zagreb, Zagreb, Croatia

5 Hochschule Bonn-Rhein-Sieg, Germany

6 Instituto Superior Técnico, University of Lisbon, Portugal

7 Sapienza University of Rome, Italy

*Address all correspondence to: matthew2.studley@uwe.ac.uk

References

[1] Lima PU, Nardi D, Kraetzschmar GK, Bischoff R, Matteucci M. RoCKIn and the European robotics league: Building on RoboCup best practices to promote robot competitions in Europe. In: RoboCup 2016: Robot World Cup. Vol. XX. Springer International Publishing; 2017. pp. 181-192

[2] Eurobarometer, Special. Public Attitudes towards Robots. European Commission. 2012 http://www.ab.gov.tr/files/ardb/evt/Public_attitudes_toward_robots_2012.pdf

[3] Lima PU, editor. RoCKIn-Benchmarking through Robot Competitions. London, UK, London, UK: IntechOpen; 2017

[4] Basiri M, Piazza E, Matteucci M, Lima P. Benchmarking functionalities of domestic service robots through scientific competitions. KI-Künstliche Intelligenz. 2019;**33**:357-367

[5] Basiri M, Pereira J, Bettencourt R, Piazza E, Fernandes E, Azevedo C, et al. Functionalities, benchmarking system and performance evaluation for a domestic service robot: People perception, people following, and pick and placing. Applied Sciences. 2022;**12**(10):4819

[6] Ferreira F, Ferri G. Marine robotics competitions: A survey. Current Robotics Reports. 2020;**1**:169-178. DOI: 10.1007/s43154-020-00022-5

[7] Ferri G, Ferreira F, Djapic V, Petillot Y, Palau M, Winfield A. The euRathlon 2015 grand challenge: The first outdoor multi-domain search and rescue robotics competition - a marine perspective. Marine Technology Society Journal. 2016;**50**(4):81-97

[8] Ferri G, Ferreira F, Djapic V. Multi-domain robotics competitions: The CMRE experience from SAUC-E to the European robotics league emergency robots. In: OCEANS 2017 - Aberdeen. 2017. pp. 1-7. DOI: 10.1109/OCEANSE.2017.8084767

[9] Nagatani K, Kiribayashi S, Okada Y, Otake K, Yoshida K, Tadokoro S, et al. Emergency response to the nuclear accident at the Fukushima Daiichi nuclear power plants using mobile rescue robots. Journal of Field Robotics. 2013;**30**(1):44-63

[10] Ferreira F, Ferri G, Petillot Y, Liu X, Franco MP, Matteucci M, et al. Scoring robotic competitions: Balancing judging promptness and meaningful performance evaluation. In: In 2018 IEEE International Conference on Autonomous Robot Systems and Competitions (ICARSC). IEEE; 2018. pp. 179-185

[11] Ferri G, Ferreira F, Faggiani A, Fabbri T. From ERL to RAMI: Expanding marine robotics competitions through virtual events. In: OCEANS 2021: San Diego – Porto. 2021. pp. 1-8. DOI: 10.23919/OCEANS44145.2021.9706098

[12] Ferri G, Ferreira F, Viguria A, Perez-Grau F. Aerial and underwater robots in competitions. IEEE Robotics & Automation Magazine. 2018;**25**(4):119-120

[13] Wang L, Iocchi L, Marrella A, Nardi D. Developing a questionnaire to evaluate customers' perception in the Smart City robotic challenge. In: 2019 28th IEEE International Conference on Robot and Human Interactive Communication (ROMAN).

IEEE; 2019. pp. 1-6. DOI: 10.1109/RO-MAN46459.2019.8956394

[14] Wang L, Iocchi L, Marrella A, Nardi D. HRI users' studies in the context of the SciRoc challenge: Some insights on gender-based differences. In: Proceedings of the 8th International Conference on Human-Agent Interaction (HAI '20). New York, NY, USA: Association for Computing Machinery; 2020. pp. 287-289. DOI: 10.1145/3406499.3418763

[15] Marella A, Wang L, Iocchi L, Nardi D. A methodology to design and evaluate HRI teaming tasks in robotic competition. Journal ACM Transaction on Human-Robot Interaction. 2022. DOI: 10.1145/3528415

[16] Antonioni E et al. Nothing about us without us: A participatory design for an inclusive signing Tiago robot. In: 2022 31st IEEE International Conference on Robot and Human Interactive Communication (RO-MAN). 2022. pp. 1614-1619. DOI: 10.1109/RO-MAN53752.2022.9900538